Team Development for
High-Tech Project Managers

For a complete listing of the *Artech House Technology Management and Professional Developement Library,* turn to the back of this book.

Team Development for
High-Tech Project Managers

James Williams

Artech House
Boston • London
www.artechhouse.com

Library of Congress Cataloging-in-Publication Data
Williams, James.
    Team development for high-tech project managers / James Williams.
       p.   cm.—(Artech House technology management and professional
      development library)
      Includes bibliographical references and index.
      ISBN 1-58053-134-2 (alk. paper)
      1. Project management.   2. Teams in the workplace.   I. Title.   II. Series.
    T56.8 .W53 2002
    658.4′02—dc21
                                          2001056565

British Library Cataloguing in Publication Data
Williams, James
    Team development for high-tech project managers. —(Artech House technology
    management and professional development library)
    1. Teams in the workplace   2. High technology industries—Management
    3. Industrial project management
    I. Title
    658.4′02
    ISBN 1-58053-134-2

Cover design by Gary Ragaglia

© 2002 ARTECH HOUSE, INC.
685 Canton Street
Norwood, MA 02062

International Standard Book Number: 1-58053-134-2
Library of Congress Catalog Card Number: 2001056565

10 9 8 7 6 5 4 3 2 1

Contents

3   Types of teams 33

4   Stages of team development 47

5   Communication systems for teams in a technical environment 63

Preface

Two heads are better than one.
There's no "I" in "TEAM."
Together everyone achieves more.

How many times have we all heard these adages? For most of us, probably more times than we care to remember. I repeat them here to make an important point, though. The very fact that so many of these sayings exist, and the fact that they are so ingrained in organizational culture, points out the importance of teamwork in today's business world. In fact, many of us probably take teamwork for granted. However, in the fast-paced high-tech business arena of today, it is essential that project managers concentrate first on building a high-performing team before they put themselves to the task of accomplishing project tasks. I have been involved with a number of high-tech organizations in various capacities over the last 15 years, and I know from experience (sometimes painful experience) that a project is doomed from the start if it is not managed with a cohesive, committed team. This book is a compilation of those years of experience supplemented by research from some of the leaders in project management, quality, and human resources. It is a no nonsense, no frills, where-the-rubber-meets-the-road text that looks at techniques that work and techniques that don't. I'm of the opinion that just because a management technique was effective 10 years ago, or even 2 years ago, doesn't mean that it's beyond review and revision. I believe it is important for us to review the processes we use to build teams and conduct projects and be open to continuously modifying our thinking to match the current business climate. In the high-tech world this is essential as the business climate changes almost daily. I welcome you to read this text and pick it apart. Feel free to take issue with the thoughts I present here. My experience has shown that we learn much more when we voice our disagreements with one another and when we engage in spirited discussion of issues that matter. The real value of this book will not necessarily be found in using this text as a cookbook to project success, but rather in using it to stimulate your thinking and helping you to view project management and team building in new ways.

My primary audience for this book is managers who are relatively new to project management and managers with experience who might want to brush up on particular topics. As I said, my intention is to stimulate your thinking and challenge assumptions that you hold to be true in the areas of project management and team development. While this is not an academic or scholarly text, those in academia (students and instructors) may find use in the ideas presented here as they spring from real-world experience. Theory is fine but I believe students of project management need exposure to how that theory plays out in real organizations. I have found that much is written about what goes right in organizations, but few people are willing to discuss their mistakes and failures. However, it is precisely those mistakes and failures that teach us the most and so I present my experiences, warts and all. Readers should take the information herein to help them avoid some of those mistakes as they make their way through team development in high-tech projects.

The text is divided into three basic sections. Chapters 1 through 4 deal with the general mechanics of team and project formation. Chapters 5 through 11 look at topics pertaining to team leaders and participants as they work through the project process. Chapters 12, 13, and 14 are miscellaneous topics of interest to team members and Chapter 15 is a look ahead at where the profession of project management is going and it's impact on team development. Of particular note is Appendix C, which ties a real-world project to most of the chapters of the text. It might be a good idea to read this first before you dive into the text and then again after you complete the text. It should give you a look at how all these topics relate to work done out in the real world. After all, that's where it counts.

1

Project-management principles

An introduction to managing technical projects and project participants

Projects and project management have been with us for much longer than the few decades we usually think of when we consider these terms. The Great Wall of China and the pyramids in Egypt are examples of projects undertaken thousands of years ago which required a good deal of project management to complete. More recent large-scale projects include the Apollo Space Program, the Manhattan Project, and the Desert Storm military action in Iraq. Projects closer to home may include installing a private intranet for the first time or leading a group of engineers in the design of a new product. Projects do not necessarily have to be on such a grand scale. Adding a deck to your house or taking a vacation trip to the beach are smaller scale projects that many of us have experienced. Whether you realized it or not, if you planned and executed such a project, you were practicing project management. Formal project management as we know it today has its roots in the military from the 1950s. The federal government, particularly the Department of Defense, realized that it needed a process with which to keep track of large-scale projects with thousands of tasks and hundreds of suppliers and subcontractors. Principles of formal project management sprouted and began to grow, spreading from the government to such private industries as construction. Through the 1970s and 1980s these principles were refined and introduced into more and more areas of the business world. Today project management is recognized as a discipline unto itself, and the role of the project manager continues to grow in importance in organizations. In order to better understand the different elements of project management, particularly the make up and the role of the project team, in the rest of this chapter I will discuss some of the basic elements and principles involved in modern formal project management.

Definition of a project

The best place to start is usually at the beginning, so let's begin by defining exactly what we mean by the term project. All projects typically have the following four elements in common:

1. *A goal.* Projects are, or at least should be, designed to achieve a particular result. This result, or goal, should be clearly and unequivocally defined. All resources consumed by the project should be utilized in achieving this goal. The goal serves as a beacon to guide the project, much as a lighthouse serves to guide ships coming in to shore or the North Star guides travelers at night. As long as you can see your goal, you know where you're headed. We have seen the importance placed on goals in the military over the last decade. Increasingly, Congress and the public want to know what the specific goal of a military action is before armed forces personnel are committed. The old axiom, If you don't know where you're going, how do you know when you get there? aptly gets at this point.

2. *Coordinated, interrelated subtasks.* The ultimate goal can be subdivided into smaller goals, each of which may be subdivided further. Eventually you end up with a list of tasks that are in small enough chunks to allow you to assign specific resources and to estimate time to completion. Each of these subtasks should be stated just as clearly and unequivocally as the ultimate goal. The tasks must also be organized so that they occur in the correct sequence. For example, when you start a car, the task of inserting the key into the ignition must occur before the task of turning the ignition to the

start position. One tool that project managers use to organize project tasks is a work-breakdown structure (Table 1.1). This is a formal outline-style document that shows how subtasks roll up toward the ultimate goal. It allows for display of estimated durations for each task and may also show costs to complete each task. Another tool to aid in task organization is the network diagram (Figure 1.1). This is a graphical representation of the flow of tasks in a project. It can be used to show the critical path through the network of tasks. The critical path is the chain of tasks which, front to back, takes the longest length of time to complete. If tasks on the critical path take longer to complete than estimated, the end date for the total project will have to be pushed out further into the future.

3. *Finite duration.* Finite duration is the characteristic that separates projects from processes. Processes are repeated over and over ad infinitum, while projects have a defined beginning and end. The end date for a project may slip, but an end date is eventually reached. In fact, much of the work of the project manager is spent in trying to keep the project headed to the agreed-upon end date and in reporting interim project status in relation to the end date. Project managers and team members should keep in mind that the end date for the project does not necessarily mean that their work on the project is done. In a production environment, a production turnover has to take place from the project team to the production support team. In some instances, some or all of the project team members are shifted onto the production support team. The end date on a project plan is a milestone, but it is not necessarily the true end date of involvement for the project team.

4. *Uniqueness.* All projects are unique. This does not mean that every project is a start-from-scratch effort. In fact, as a project manager gains experience, he or she will learn to rely on past projects for building blocks to insert into the current project. Some elements will be new, though. In a construction environment where a contractor may build numerous houses from the same set of blueprints, soil conditions may vary, materials may vary, and weather conditions may vary. In a computer application development

TABLE 1.1   A Work-Breakdown Structure

Project statement: Move the IT programming staff from the first floor to the third floor in two months at a cost not to exceed $100,000.

ID	TASK NAME	DURATION (DAYS)	START DATE (AT 8:00 A.M.)	FINISH DATE (AT 17:00 P.M.)	PREDECESSORS
1	Office layouts drawn	17	7/13/99	8/04/99	—
1.1	Relationship charts prepared	10	7/13/99	7/26/99	—
1.1.1	Interviews conducted	5	7/13/99	7/19/99	—
1.1.2	Charts prepared	5	7/20/99	7/26/99	3
1.2	Block layouts drawn	3	7/27/99	7/29/99	4
1.3	Detail layouts drawn	4	7/30/99	8/04/99	5
2	Office equipment identified	2	7/13/99	7/14/99	—
2.1	Equipment to keep identified	2	7/13/99	7/14/99	—
2.2	Equipment to discard identified	2	7/13/99	7/14/99	—
2.3	Equipment vendor contacted	1	7/13/99	7/13/99	—
3	Office area prepared	42	7/15/99	9/10/99	7
3.1	Electrical service installed	14	7/15/99	8/03/99	—
3.2	Telephone service installed	14	8/04/99	8/23/99	12
3.3	Computer links installed	14	8/24/99	9/10/99	13
4	Office moved	4	9/13/99	9/16/99	11
4.1	Equipment moved	2	9/13/99	9/14/99	—
4.2	Equipment installed	2	9/15/99	9/16/99	16
4.3	Personal materials moved	1	9/13/99	9/13/99	—

environment, a developer may rely on previously developed modules of code, but the way in which those modules are put together may vary, user requirements will certainly vary, and the computing environment itself may be different than when those modules were used before. The uniqueness of projects means that each

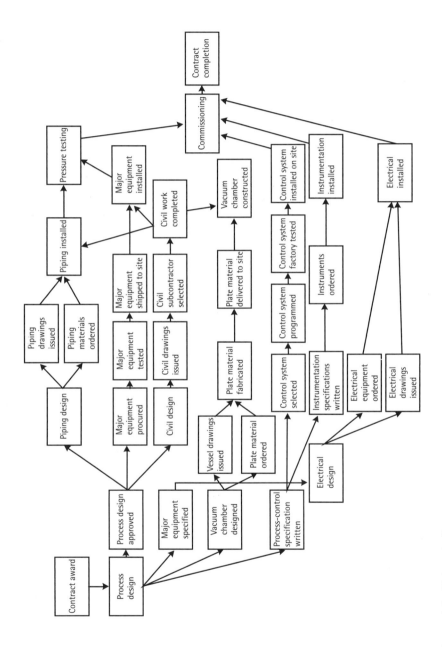

Figure 1.1 Network diagram.

project must be managed as an individual entity. In fact, it is the uniqueness of projects that forces us to employ a project management process in the first place. If we always could perform the same tasks the same way, we would always know how long a project would take and we would always know just how to perform the tasks. Since these elements vary from project to project, we need to employ a plan so as to maximize the efficiency of our efforts.

Definition of project management

The Project Management Institute defines project management as "...the application of knowledge, skills, tools, and techniques to project activities in order to meet or exceed stakeholder needs and expectations from a project. Meeting or exceeding stakeholder needs and expectations invariably involves balancing competing demands among scope, time, cost, quality, and stakeholders with differing needs and expectations, identified requirements, and unidentified requirements." David I. Cleland, in his book *Project Management: Strategic Design and Implementation,* defines it as "...the art of directing and coordinating human and material resources throughout the life of a project by using modern management techniques to achieve predetermined objectives of scope, cost, time, quality, and participant satisfaction" [1]. In layman's terms, project management concerns getting the job done on time, within budget, and to specifications. It is the sum total of all processes used on a project to bring it to fruition within certain boundaries. You'll notice the recurring themes of time, cost, and specs in all these definitions. This is no coincidence. Those three elements taken together comprise the triple constraints (Figure 1.2). The whole framework of formal project management is concerned with balancing those

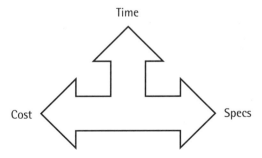

Figure 1.2 The triple constraints.

constraints to the satisfaction of the sponsors, stakeholders, and supporters. Any change in one of the constraints has a direct impact on the others. For example, if you shorten the timeline for the project, cost will go up because more resources will have to be dedicated to the project in order to accomplish the tasks. If new specifications are added, new tasks will be added to the project plan, thus lengthening the timeline of the project and increasing the cost. If cost overruns are encountered on certain tasks, specifications may be changed in order to bring total project cost back in line with the established budget.

Definition of a team

We can agree that a team is a group of individuals organized for a particular purpose. A basketball team, for example, is comprised of players and coaches for the purpose of playing and, hopefully, winning basketball games. Now consider this: Are the equipment managers part of the team? How about the people who keep scores and statistics for the team? What about the person who maintains the playing facility? Certainly all of the aforementioned people contribute to the process of staging and executing a basketball game, but are they part of the basketball team? Let's look at three team aspects that will help us more clearly define what we mean by the word team.

1. *Purpose.* Teams are organized for a purpose. They may be assigned a process to perform or a project to accomplish. Teams often have a mission statement to help guide their actions and focus their energies. The emergence of the purpose or goal is usually the spark that drives the formation of a team in the first place.

2. *Duration.* Teams may be chartered for long-term tasks, or their existence may be only temporary. Whether the team is permanent or temporary, its existence is linked to the task it is assigned and just about every task that is assigned to a team has a duration. The duration may be stated explicitly (e.g., for four months), or implicitly (e.g., until the product line ceases to achieve a predetermined level of profitability). The last example may sound like an unlimited duration, but reality dictates that products will eventually become obsolete; therefore, what at first sounds like an unlimited amount of time becomes only a very long, finite amount of time.

3. *Member function.* The members of a team may bring similar abilities to the team or they may be drawn from a wide range of experiences. The purpose of the team will dictate what skills are needed from the members. In either case, each member will have a function to perform that will contribute to the achievement of the overall goal of the team and each member will depend on all the others to help accomplish the team goal. Sometimes members may perform the same general function, but have specialties within that function. To return to our basketball example, all the players are expected to dribble, pass, rebound, and shoot, but the point guard is expected to pass more than rebound and the center is expected to rebound more than pass.

Now that we have looked at three aspects of a team, the question remains: What distinguishes a team from just a group of people? Purpose is the overarching characteristic that separates a team from a group. All the managers of a corporation, for example, comprise a group, but they can't be considered a team because they lack a clearly defined single purpose. They don't depend on each other to accomplish a team goal. They are a group with common general interests, such as corporate profitability and customer satisfaction, but they have individual goals to achieve as well. There is no particular duration to the group and the functions of group members are not necessarily linked in any way. Teams, on the other hand, have a purpose and the members exhibit a high degree of interdependence as they strive to accomplish that purpose.

Role of the project manager

The project manager must be a statistician, a planner, a human resource expert, a budget whiz, and a politician—sometimes all at once! The project manager must be able to perform the following tasks:

- ◆ Organize resources in order to achieve the goals of the project.

- ◆ Provide leadership for the project team. Foster teamwork among team members and resolve conflicts before they impact project progress adversely. Mentor the members of the team and at times wear a coach's hat, spurring the team on to greater productivity;

- Evaluate project progress so as to determine whether the project is on schedule from both a budgetary and a time standpoint. Must be able to track project progress and demonstrate statistically where the project stands in relation to the project plan;

- Manage the project in the political context in which it is being performed;

- Report project status periodically to sponsors, stakeholders, and supporters. Present good news and bad news with equal aplomb.

I like to think of the project manager as the control point for the project. All project activities hinge on this one person. To use an analogy, think of the project manager as a stagecoach driver in the old West. The team of horses that pulls the stagecoach represents the project team. The sponsors, stakeholders, and supporters are the passengers inside the stagecoach. The horses don't know when to begin pulling, when to stop, or where to go unless they receive leadership and direction from the driver. The passengers can't reach their destination unless they work through the driver and give clear directions as to where it is that they want to go. I think this illustration serves us well in helping to visualize how the project manager helps make the connection between the project team and the outside world.

The project life cycle

As stated earlier, projects have a definite beginning and a finite duration. Like a good story, they have a beginning, a middle, and an end. According to the Project Management Institute, a project has four phases: concept, definition, execution, and finalization. Another way to categorize project phases is to break them into needs recognition, project selection, project planning, project implementation, project control, evaluation, and finally completion. Keep in mind that feedback loops exist between the phases. For example, the evaluation phase may uncover elements within the project that need to be taken back to the planning phase and then run forward through the process again. Any way you categorize the phases of a project, each phase should have a deliverable and a decision should be made as to whether the project should continue to the next phase. This structure helps in planning the overall project by predefining a subset of the overall deliverables. Thus, the project team and the stakeholders, sponsors, and supporters have an idea of what to expect before the details of the project are ever determined. The project life cycle structure may also help determine what resources are necessary for each phase, another aid to overall project planning. Finally, the type of project and the industry in which the project is being performed may dictate what the project life cycle looks like. A construction project life cycle may use a highly detailed waterfall methodology (Figure 1.3) to organize its projects, whereas a software development project life cycle may use an iterative, rapid application development model (Figure 1.4). Where the construction project may be very regimented and have much historical data behind it from other projects, the software development project may employ more of a trial-and-error approach with little historical project data to rely on. Suffice it to say that one project life cycle size does not fit all, but some project life cycle should be employed as part of the overall project planning process.

Project management processes

Projects are comprised of interrelated processes. The project manager must manage these processes so that efficiency in one process is not bought at the expense of the efficiency of another process. There are generally five process groups for any project:

1. *Initiation.* This involves needs recognition and a commitment to begin a project.

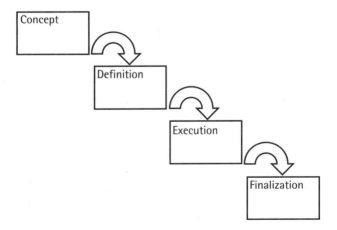

Figure 1.3 Waterfall methodology: Steps are completed sequentially and tasks are single-threaded through the project process.

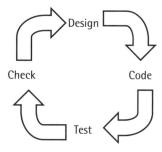

Figure 1.4 Rapid application development: Steps are performed iteratively until the project is completed.

2. *Planning.* This group encompasses the development of a workable plan to accomplish the goal of the project.

3. *Execution.* Execution processes are concerned with organizing resources to actually get the project done within the bounds of the triple constraints.

4. *Control.* Controls are necessary to ensure that objectives and deliverables are being met as dictated by the project schedule.

5. *Closure.* Every project should be formally closed with signoff from all sponsors, stakeholders, and supporters.

Each process group produces output that becomes the input for the next group. It may also happen that process phases overlap or are repeated. The process groups can be further subdivided into discrete processes. The Project Management Institute organizes these processes in the following manner:

Initiation

1. *Initiation:* a commitment to begin the project.

Planning

1. *Scope planning:* development of a scope statement;

2. *Scope definition:* division of deliverables into manageable pieces;

3. *Activity definition:* identification of specific work packages to be executed;

4. *Activity sequencing:* identification of work package interaction;

5. *Activity duration estimating:* identification of work package durations;

6. *Schedule development:* creation of the project schedule;

7. *Resource planning:* assignment of resources to work packages;

8. *Cost estimation:* development of work package costs;

9. *Cost budgeting:* allocation of cost estimates to work teams;

10. *Project plan development:* compilation of previous planning processes.

Execution

1. *Execution:* performance of the project plan;

2. *Scope verification:* acceptance of the project scope;

3. *Quality assurance:* periodic evaluation of project performance;

4. *Team development:* develop of team member skills to enhance project performance;

5. *Information distribution:* disseminating information to sponsors, stakeholders, and supporters;

6. *Solicitation:* obtaining quotes, bids, offers, and proposal as necessary;

7. *Vendor selection:* selection of third-party resource providers;

8. *Contract administration:* management of third-party vendor relations.

Control

1. *Change control:* coordination of project change requests;

2. *Scope control:* coordination of modifications to project scope;

3. *Schedule control:* coordination of modifications to project schedule;

4. *Cost control:* containment of project costs to budgeted levels;

5. *Quality control:* monitoring project results as compared to project specifications;

6. *Performance reporting:* collection and distribution of project performance information;

7. *Risk response control:* containment of project risk as defined by the risk management plan.

Closure

1. *Administrative closure:* gathering and distributing formal closure material to sponsors, stakeholders, and supporters;

2. *Contract close-out:* completion and settlement of any contracts opened for the project.

In addition to these processes, there are facilitating processes that occur across all these processes. They include, but are not limited to, quality planning, organizational planning, communication planning, solicitation planning, staff acquisition, and risk identification, quantification, and response development.

Conclusion

As a wrap up to this whirlwind tour of project management, always remember that the goal of a project manager is to get the work done within the triple constraints. Keep the following points in mind and your journey to the project's goal will be much smoother than it otherwise might be:

1. *Do your homework.* Keep yourself apprised of the most current best practices in project management. Don't just fly by the seat of your pants. You're certain to fail if you take this approach.

2. *Sweat the details or bleed the implementation.* It is human nature to want to immediately rush to the end goal of a journey. As a project manager you must fight the urge to ignore the upfront details of planning just to get to the execution stage of the project. Do the

right things right the first time and you will reap the dividends by the end of your project.

3. *Plan for the unexpected.* Risk management must be performed over the entire life of the project. Risk cannot be completely avoided. The unexpected will occur, but a good project manager can mitigate the effects of unforeseen problems by following good risk identification, quantification, and mitigation strategy development techniques.

4. *Take nothing at face value.* Make sure that you as a project manager take the time to perform root-cause analysis when problems arise so that when fixes are applied they are appropriate. Question the motives and desires of your clients ad infinitum. Be sure that you know exactly who are the true sponsors, stakeholders, and supporters of a project and what it is that they want. There is nothing more discouraging than to deliver exactly what your customer asked for only to find that it wasn't what they wanted.

5. *Maintain flexibility.* Change is a fact of life in project management. A certain degree of formality is necessary to manage a project, particularly a large-scale project, but it pays to be nimble. Don't get locked into any process or system so tightly that you can't move with the flow of your clients' needs. Remember to bend and not break.

Reference

[1]   Cleland, D. I., *Project Management: Strategic Design and Implementation,* New York: McGraw-Hill, 1994, p. 5.

2

Project-requirements development

Introduction

Now that we have some of the basics of project management squared away, we should take some time to discuss project selection and requirements development. After all, it is from our organization's projects that project teams spring. Let's take a look now at how projects come to be.

Project selection

Project ideas come from various sources and are driven by both external and internal forces. Competition, new market opportunities, customer demands, and government regulations all may be project drivers. In some cases these external forces dictate that an organization undertake a project, whether the organization is really ready for the project or not. The problem most organizations face is that they spend most of their time in a reactive mode when dealing with project drivers. Organizations should strive to be more proactive in project generation and selection. They need to do a better job of analyzing the market, their customers, and the regulatory environment so that they can utilize the project generation and selection techniques that will be discussed later in this chapter.

Try this simple exercise. Take five minutes and jot down as many project improvement ideas as you can for your organization. My expectation is that you will be able to develop a fairly substantial list. Consider if the members of your immediate work group performed this same exercise and then you all compiled your lists, combining duplicate project ideas. This document would be quite a substantial list of potential projects for your organization to undertake. Now take this idea even further. Consider if everyone in the organization performed this exercise. Your organization would have a huge list of projects that potentially could be undertaken. Some might be cost-saving measures, some might be new product ideas, and some might be process improvements. The question then becomes, how do you choose which projects to actually work on and how do you prioritize those chosen projects? What factors come into play when selecting projects for the organization? Why do seemingly worthwhile projects go by the wayside while other, apparently less worthy projects, receive resources to be worked? How do you generate project ideas in an orderly fashion in the first place? Organizations need a formal methodology to follow for project generation and selection. Depending on the types of projects typically worked in a particular organization, a number of methodologies may be needed. Several methodologies will be discussed here which deal with different types of projects as well as several idea generation techniques.

Let's first discuss some project idea generation techniques. You probably have already used some of these techniques, but the real skill involved in idea generation is to know which technique to use in which situation. The following are some suggested techniques:

- *Brainstorming.* Brainstorming is a technique with which you seek to glean ideas from every member of your project team, regardless of the practicality of the idea. Each member of the group is asked to propose an idea in turn. If a member does not have an idea at that particular moment, he or she may pass their turn. Group members are encouraged to use the ideas of their fellow group members, which necessitates the members being either in the same room for the brainstorming session or at least connected through some means of telecommunications. All ideas are recorded at this point. No idea is evaluated or discarded at this time. The point of this stage of the process is to generate as many ideas as possible and to encourage team members to exercise their creative thinking to the maximum. Once all team members have exhausted their ideas, the team can move to step two of the process. This involves evaluating the merits of each idea, discarding some and refining others. The end product of the whole brainstorming process should be a list of ideas that the group is prepared to carry forward into the project selection process.

* *Delphi technique.* Delphi is a technique whereby an analyst polls a group of subject-matter experts for ideas. The ideas should include some brief detail or at least an explanation of why the expert feels this idea would be viable as a project. The analyst reviews all the responses, summarizes the unique responses and combines the like responses. This list is then put back out to the subject-matter experts with a request for further evaluation and substantiation of each idea. The process of polling the experts may be repeated several times until the analyst has a list of unique, well-substantiated project ideas. This list is then prepared by the analyst for submission into the project selection process.

* *Nominal group technique.* The nominal group technique is similar to brainstorming, but rather than generating ideas at random, the facilitator for the group begins with a seed proposition. The group members then work off that seed idea in turn, listing the ideas in written form as they go. Once the group members have exhausted their ideas they evaluate the total list, rank the items, and select a certain portion of the list for project selection.

* *Crawford slip.* The Crawford slip process again involves a facilitator beginning the process with a seed idea. In this technique the idea is formed as a question. Group members each write one response to the question on a slip of paper. The group members are then asked the same question again, worded slightly differently. Again they write their answers on slips of paper, but this time the answers must be different from the first. This process is repeated 10 times. The answers are then examined with like answers being combined. In this manner, a list of project ideas is generated, but the team members do not feed off each other. The ideas are strictly their own.

All these techniques have been proven to be successful in generating ideas. The technique you choose for a particular situation will depend on a

3. Establish a matrix of criteria against which to evaluate the proposals.

4. Incorporate any visual representations of the proposals into the material to be considered.

5. Validate the criteria developed to this point.

6. Choose the criteria that will be used for proposal comparison and discard any irrelevant criteria.

7. Choose a reference proposal against which to begin comparison.

8. Choose a scoring legend.

9. Compare each proposal on each criterion.

10. Tally and evaluate the scores of each proposal.

11. Examine the highest scoring proposals for weak scores on any one criterion. If the weakness can be corrected, introduce a new corrected proposal into the mix.

12. Examine the lowest scoring proposals for weak scores on any one criterion. If the weakness can be corrected, introduce a new corrected proposal into the mix.

13. Eliminate weak proposals that cannot be strengthened.

14. If truly strong proposals do not emerge, examine the comparison criteria for ambiguity. Also, examine proposals to see if any can be combined. If criteria are ambiguous, correct and start the process over. Combine proposals that can be combined.

15. Reprocess the matrix using the strongest proposal from the first pass as the reference proposal.

You will notice that quantitative results are not part of the Pugh methodology. This methodology relies on the iterative nature of the process and the continual strengthening of the proposals to produce one or more projects that are strong uniformly across all comparison criteria. You will also notice that as successive iterations of the process are performed, the reference proposal is the strongest proposal from the previous iteration. Thus, the bar is raised, so to speak, until the project selection team is satisfied that the strongest proposal has emerged and any weaknesses in that particular proposal have been eliminated. One of the greatest strengths of the Pugh method is that it allows for new concepts to be introduced into the process once it has begun, which allows for a greater chance that alternative strategies not considered at first may arise. This is where the idea of divergent thinking comes into play in that project ideas that diverge from the original list are allowed to be introduced. These divergent ideas then have the opportunity to rise to the top of the list, thus providing a means for breakthrough thinking to come into play in this project selection method. The main downside to the Pugh methodology is that it is very time consuming as weak project proposals receive quite a bit of attention before they are eliminated from consideration.

The two methodologies presented so far both have a requirement for developing some kind of weighting system to rank criteria, and thus, to arrive at some quantitative ranking of project proposals. A methodology developed by Tzvi Raz, a professor at Tel Aviv University, eliminates the need for such a weighting system, while still yielding a quantitative method of selecting projects [3]. The Raz method makes use of statistical analysis, specifically standard deviation and the coefficient of variation. It requires that a table of criteria, each of which can be represented in a quantitative manner, initially be established. The criteria may be such measurable elements as number of people required to perform a certain phase of the project, dollar amount needed to begin the project, estimated hours to complete the project, and others. The following five steps are then performed iteratively until one proposal emerges as the strongest:

1. Calculate the standard deviation and the coefficient of variation for each set of values assigned to each element of the matrix of criteria (Box 2.1).

2. Rank the criteria in descending order of the coefficient of variation (Box 2.1).

3. Eliminate criteria from the matrix that fall below an agreed-upon level of the coefficient of variation. This cutoff may be predetermined, but it is usually better to set the cutoff based on natural gaps in the coefficients of variation.

4. Eliminate projects from further consideration that are dominated by all other alternatives. For example, if a project scores lower in virtually every category than all other projects, it should be

eliminated from further consideration. It would be inconceivable that this project, under any weighting of the criteria, would rise to the top of the selection list.

5.   Examine the remaining projects to determine if a clear alternative is present. If not, continue to repeat the process with the remaining criteria and projects until a clear alternative emerges. A clear alternative would be one that dominates all the others in virtually all criteria.

One advantage of this method is that it does not require weights to be assigned to each criterion. Hard, verifiable numbers are used, thereby eliminating any bias that might be introduced by the subjective assignment of a score to a criterion. Another advantage is that attention is focused on the criteria that really differentiate between the projects, rather than on criteria that has little impact on the selection process. Finally, the selection process becomes simpler with each iteration, rather than remaining as complex, or possibly becoming more complex. The main drawback to this methodology is that it requires criteria that can be expressed in a quantitative manner. Qualitative data can be converted to quantitative data, of course, but then this diminishes the quantitative advantages of the method.

In general, any project selection methodology should be linked to the overall goals and objectives of the organization. No project should be selected that does not, in some way, further the stated objectives of the organization. High-performance organizations establish overall corporate goals and then flow these goals down through the organization via establishment of personal employee objectives that link back to the corporate objectives. These goals may be modified as the business climate changes, but, nonetheless, they exist and should be distributed to all employees. The aim here is to provide total focus for the organization and to get everyone pulling in the same direction. Of course, we all can name projects within our organizations that do not appear to link to any particular organizational goals. We have to ask ourselves how this can happen if we are using the principles just described here. There may be a number of reasons, and these are pitfalls that you as a project manger should be aware of and try to avoid. First, it may be that the goals and objectives of the organization have not been clearly stated or have not been pushed down into the front lines of

Box 2.1   Standard Deviation and Coefficient of Variation

For a data set containing n measurements the standard deviation, *s*, is defined as

$$s = \sqrt{\frac{\sum\limits_{i-1}^{n}\left(x_i - \underline{x}\right)^2}{n-1}}$$

where x_i is the value of an individual measurement and x is the average value of the measurements. The standard deviation is a measure of the precision of the measurements. If *s* is small, the measurements are tightly clustered around the average, and the precision is high. If *s* is large, the individual measurements exhibit a wide range of values and the precision is low.

The coefficient of variation can be expressed as follows:

$$CV = (S/X) \times 100$$

where *s* = standard deviation of a sample and *x* = the average value of the measurements of the sample. The coefficient of variation is always expressed as a percentage, therefore standard deviation divided by the mean is multiplied by 100. It is used to show variation in a sample relative to the mean and is a useful tool when comparing two samples.

the organization. This is unlikely if you are operating inside a high-performance organization, but the chance is there, however slim.

Another more likely culprit would be political machinations within the organization. I have seen firsthand how politics inside the organization can lead you to project selection ruin. A large telecommunications firm at which I was employed selected a project that was designed to revamp the way requests that came into our external security department were entered and stored. While this project certainly promised to reduce the workload of the two clerks who manually performed this duty, the overall benefit to the organization was marginal. The driver behind the project was the desire of a particular IT manager who wanted to showcase the talents of her staff on a particular platform. This manager was in charge of a new development group that was working with newer, more cutting-edge technology. She was able to convince our director that this security project would be an excellent opportunity to increase productivity in the security department, thus freeing the clerks to perform other tasks. The director bought into the proposal and the project was on. At this point you should be asking yourself, where was the formal idea generation methodology and where was the formal project selection methodology in this scenario. My point with this example is to illustrate that neither process came into play. The project was selected solely on the perceived merits of the particular IT manager. The project was worked and a product was produced; however, the software developed was so resource intensive the desktop machines used by the clerks could not handle it. The project consumed more resources overall than it saved and eventually the software was shelved. Not only were the dubious merits of the project not realized, scarce company resources were consumed all due to the political contriving of a manager. Let this be a lesson in the perils of abandoning or circumventing a formal project selection methodology.

Project initiation

Project idea generation and project selection are truly the first steps in project initiation, but given the organizational structure of most companies, a project manager and a project team may not be assigned until after these steps are completed. For this reason, it makes sense to discuss project initiation, as most of us understand it, at this point.

For purposes of the immediate discussion we will assume that a project manager has not been assigned at this point. Let me make it clear, though, that this is not the ideal situation. Ideally we would like our organizations to be more "projectized" and have a project manager involved during both idea generation and project selection. This allows the organization to make use of the project manager's wealth of project experience as early in the project process as possible. It also gives the project manager the opportunity to garner as much information about the project as possible. It allows him or her to be involved in the discussions and the brainstorming that led to the project idea in the first place. I know from personal experience that this is of great benefit. My experience in the IT field has been that we are only brought into projects when the client departments first realize they will need IT resources. Oftentimes a project will have three months of client history behind it before it is turned over to an IT project leader to shepherd through to fruition. Many times in this situation decisions will be made by personnel inexperienced in project management which only complicate the job of the project manager once one is appointed to the project. I had the good fortune to participate on a project in which the project manager participated from the idea generation stage. Meetings were conducted and decisions were guided by an experienced project manager who was able to keep the team on track from day one. The clients had a formal project structure through which to work rather than just being left to freehand the effort and potentially get the project headed off down the wrong road. Despite the fact that the project involved a complex system replacement, we were able to bring it in on time, within budget, and to initial specifications. As a contrast, I am currently involved in a project in which no project manager was named initially. A group of clients developed the project idea and even went as far as to draw up design specifications. The project was due six months ago and the end is not yet in sight. The project manager has spent the majority of his time reworking specs, digging out specs that were overlooked, and trying to coordinate legitimate testing scenarios. The difference between early and late involvement by a project manager and the application of sound project management techniques is like night and day.

What do you do then when the project manager selection is not made until this point in the process? All is not lost, but it is imperative that the project manager takes control quickly and decisively. The project manager must first acquaint himself or herself with the details of the task that is to be accomplished. Face-to-face meetings with the stakeholders are a good idea at this point so that the project manager may begin to understand exactly what the stakeholders expect from the project. This is also the time to begin managing customer expectations. A project manager that fails to manage customer expectations from the early stages of a project will soon find that no matter how well things are going, the customers are never happy. Stakeholder expectations can be translated into project goals and help to insure that the project is headed in the direction the clients intend, regardless of what they officially have asked for. The other major task for the project manager at this point is to assemble the project team. Team members may be assigned to the project already with no input from the project manager or the project manager may be fortunate enough to have some say in picking the team members. Again, the project manager may be able to have some influence, even if team members have already been assigned, if he or she uses the face-to-face stakeholder meetings effectively. The project manager may be able to capture additional strategic team members, or he or she may be able to exchange current team members for members who might be of more value. The very minimum the project lead should accept at this point is a commitment from the stakeholders to provide client subject matter experts as needed throughout the life of the project. The overall goal of this phase of the process is to open the lines of communication between the project manager and the stakeholders and prime the pump to get information flowing between the stakeholders and the project team. Once the team is assembled and the lines of communication have been established, the project team can move on to project requirements definition.

Requirements-definition cycle

Just as it is necessary to implement formal processes for project idea generation and project selection, so it is necessary to follow a formal process for project requirements definition. Haphazard requirements definition will always add unnecessary time and cost to a project and will certainly reduce client satisfaction with the finished product. According to Ivy Hooks, president and CEO of Compliance Automation, Inc., studies at NASA show that cost overruns of 100% to 200% are common on projects that spend 5% or less of project costs on up-front requirements definition [4]. Projects that spend as little as 10% of up-front costs on requirements definition show overruns of only 50% or less. Minor increases in time on requirements definition seem to pay large dividends.

The requirements definition phase of the project life cycle always seems to be the one that no one really wants to tackle, and then they try to get it over with as quickly as possible. Most of the people writing requirements have no training in what to do and very few good examples to follow. The task is complicated further by the fact that the people writing requirements usually have little or no direction under which to work. Actually, requirements definition may be the most important phase of the project life cycle and is probably the phase that should receive the most time and attention. The reason no one is ever eager to jump into this phase is that it requires some hard work and it requires the clients to make tough decisions. The initial euphoria of project initiation has come to an end and the cold, hard reality of production has begun. The pressure of actually defin-ing the product or service that is desired has begun to build. This pressure can be eased by following a structured requirements definition process and by using some of the tools and techniques that will be described as we pro-ceed with this chapter.

There are six basic phases of requirements definition (Figure 2.1).

1. *Needs emergence.* Obviously, the needs for a new product or serv-ice, or the needs for updates to an existing product or service must present themselves before further action can be taken.

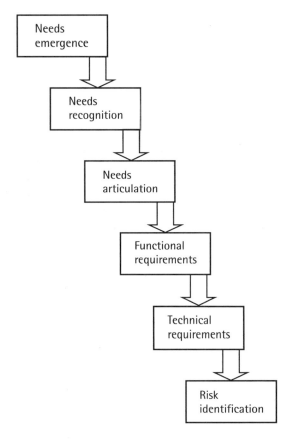

Figure 2.1 Six phases of the requirement-definition cycle.

2. *Needs recognition.* Needs must be recognized before they can be acted upon. Sometimes one set of needs feeds off another set. For example, there was no need for electric light switches until electric power generation and distribution became a reality of modern life.

3. *Needs articulation.* This is where we begin to get to the meat of the requirements definition cycle. Even though a need exists, it must be clearly articulated before it can be acted upon. This is also where many, many projects begin to go wrong. Clients may recognize a need but may not be able to articulate it clearly. They may recognize a need, but articulate requirements for another need. It is also important that the client who truly has a need to be fulfilled is the one who articulates the need. My IT experience has shown me that in many cases, client departments will recognize an emerging need but will not be willing to articulate those needs. They will turn to the IT group to articulate the needs, either because they don't want to be bothered with this task or because they feel inadequate to perform this task. The IT group then puts the need into words as they understand it, and very rarely is it expressed as the client would have expressed it. Formal requirements are developed from this initial articulation and in the end the finished product or service is less than 100% satisfactory to the client. Many organizations attempt to avoid the aforementioned situation by creating a group of functional clients to work as a go-between between the production functional clients and the IT staff. Typically, this go-between group is tasked with articulating needs and developing requirements as recognized by the production clients. This situation is little better than asking the IT staff to articulate needs because the group that truly is experiencing the need is still asking another group to articulate those needs for it. Management must realize the importance of getting the requirements definition cycle off to a good start if more of our projects are to be successful in meeting the constraints of time, scope, and budget. Day-to-day production clients and technical staff must be trained in the methods of needs articulation and definition so that as needs emerge and are recognized, they can be articulated more quickly and more accurately.

4. *Functional requirements.* Functional requirements represent detailed instructions as to how business processes are supposed to operate. The premise this statement rests on is that someone actually understands how a particular business process is supposed to operate. Again, drawing from my experiences in IT, this statement is rarely true. Clients often do not understand their own business processes, either because they are new to their work group, or they have never been given the proper training on the process, or in some cases, because they just don't care. It is not unusual for the technical support staff to understand particular business processes better than their client counterparts. In order to develop solid functional requirements, the current business process must be diagrammed. There are a number of desktop software products on the market that are capable of handling this task. The improvements or additions of the current project should then be added on top of the current business process diagram. Having a visual representation of the process in addition to a written representation increases the likelihood that all parties involved will understand the total process. The same case exists for functional requirements as existed for needs articulation—the group that has the need should compile the requirements. Management should take the time and put forth the resources to make sure that all functional groups have people in place who are capable of developing functional requirements. In addition, this is not a task that should be delegated to the technical support staff. Functional requirements are just that—functional. Technical concerns should not even be considered at this point. The thrust of this step in the process must be defining what the business needs to operate. Too often, the technical support staff becomes involved at this point and they have a tendency to shape functional requirements based on technology that is either currently in place or technology that they would like to work with. Technology should be of no concern in functional requirements development.

5. *Technical requirements.* Once functional requirements have been developed, technical specifications must be formed which will meet the business needs as spelled out in the functional requirements. Technical specs should be developed by the technicians

who will transform the specs from theory to reality. Many organizations make use of technical analysts who develop specs, which are then passed down to lower level techs for implementation. In theory this should be a smart way to make the best use of your resources. In reality this method breaks down quickly. I have found from experience that it is virtually impossible to separate the process of designing technical requirements from the process of actually transforming those requirements into a real product. Important details are invariably missed which leads to rework, sometimes forcing you to go all the way back to square one and start over again. I have worked a number of projects in the past in which technical specifications were written by an analyst and then passed to me for implementation. I received specifications that ultimately turned out to be incomplete at best, and in some cases, completely wrong. A much more efficient scenario is when the functional client writes their own specifications, then passes those specs to the technician who will be charged with developing and implementing the technical solution. Rather than involving a minimum of four people (or groups of people), only two people (or groups of people) will be involved in the communication link between the functional clients and their technical support staff. This reduces the potential communications paths from six to two, thus making communication between the groups simpler, faster, more efficient, and less prone to misunderstanding.

6. *Risk identification.* Risk identification, and from a broader perspective, project task development, is a subject on which much has already been written. I mention it here in the requirements development section because potential project risks often drive certain functional and technical requirements. For example, a functional requirement for a particular amount of uptime for a system may drive a technical requirement of redundancy within the system. The risk would be that the primary elements of the system might fail, thus the technical requirement for redundant elements in the system. I want to emphasize that it is never too early to begin risk identification, quantification, and assessment. The earlier and more thorough you can be with risk identification, the more chance you will have to work risk into requirements definition,

thereby reducing the need for risk mitigation strategies later in the project. It is always better to head risks off if you can, rather than wait for them to occur before you address them. See Appendix A for a risk and opportunity assessment framework.

Pitfalls of requirements definition

We have already discussed some of the pitfalls of the requirements definition phase of project initiation. Too few resources, both human and otherwise, devoted to requirements definition is a certain recipe for project disaster. A lack of clear project direction from the functional community leads to poor requirements definition and ultimately to poor project execution. An attempt to separate the requirements definition function from the actual project and production execution function will lead to poor requirements. In order to avoid these and other pitfalls, I recommend considering the following points:

- *Plan early and plan often.* The earlier you plan, the more chance you have to positively affect execution.

- *Be as detailed as possible with requirements definitions.* The earlier you are able to divide a project into small, manageable subtasks, the more likely you will be able to attack the project as a whole. Small subtasks are more easily digested, it is easier to predict their resource requirements, and it is easier to work small subtasks in parallel than it is to work large tasks in parallel.

- *Pay close attention to detail.* The earlier you focus on details, particularly during the requirements-definition stage, the less uncertainty you will encounter during subsequent stages.

Conclusion

The thought process required for requirements definition is difficult. It is an attempt to arrange a complex array of ideas in a logical, understandable format. The planning environment must be as stress-free as possible and the time devoted to it must be adequate to allow for deliberation, pondering, and false starts. Personnel involved in developing requirements should be relieved, temporarily, of daily activities so that they might concentrate all their energies on the planning process. The project leader must have strong convictions that the project being planned is worthwhile. Lukewarm conviction on the part of the project leader leads to lack of enthusiasm and lack of effort on the part of the project team. Finally, the more time that elapses between requirements definition and project implementation, the less chance there is that the project will be successful. This stems from the fact that (1) the farther in the past requirements definition is, the less clear the intentions of those requirements are at execution time, and (2) a long period of time between requirements definition and execution may indicate that requirements weren't specific enough. Detailed, clear, unambiguous requirements should lead to shorter development time, and thus quicker implementation.

References

[1]   Ulrich, K., and S. Eppinger, *Product Design and Development,* New York: McGraw-Hill, 1995.

[2]   Pugh, S., *Total Design,* Wokinham, England: Addison-Wesley, 1991.

[3]   Raz, T., *An Iterative Screening Methodology for Selecting Project Alternatives,* Newtown Square, PA: Project Management Institute, 1997.

[4]   Hooks, I., *Don't Start the Building Without the Blueprint: Defining the Scope of a Project,* Newton Square, PA: Project Management Institute, 1994.

3

Types of teams

Introduction

We now have a basic understanding of how projects develop, from realization of a need to requirements to project selection. In order to move forward with a project, a project team must be assembled. We'll take a look in this chapter at what teams are, how they are formed, and the types of teams that are developed to handle different situations.

What is a team?

A team can be generally defined as some number of people with complimentary skills who are committed to a common purpose and who are working both independently and interdependently to achieve particular results. The team members are held mutually accountable for the success or failure in reaching the goals set forth for the team.

Teams in organizations come in a variety of shapes and sizes. Teams may be composed of individuals who have worked together for years and know each other intimately, or they may be composed of individuals who have never met. Team size may range from a minimum of two people to a maximum of thousands. Team members may be colocated or they may be scattered across the globe. The point is, there is no neat and tidy picture of an organizational team. The needs of the project and the resources available to the organization drive how teams are structured.

Certain teams may be classified as high-performance teams. These teams seem to be able to get the job done no matter what the assignment. High-performance teams do not happen by accident. You can judge a high-performance team by the following characteristics:

- Commitment to a common goal;

- Clear roles and responsibilities;

- Informal team atmosphere;

- Contribution by all members;

- Civilized disagreement;

- Consensus decisions;

- Mutual accountability;

- Open, frank communication;

- Active listening;

- Leadership exhibited by all team members;

- Good external relations;

- Diversity of team member style;

- Self assessment by team members;

- Appreciation of both individual and team efforts.

Not only can you judge a team by these criteria; you can also develop team goals to work toward so that your team can become a high-performance team. In addition, these criteria give you a road map to use so that your team can work through the team development stages of forming, storming, norming, and performing. These stages will be discussed in detail in Chapter 4.

Another variable that contributes to team performance is team size. A number of misconceptions exist concerning the matter of team size, such as the following:

- The more team members, the more ideas;

- The more important the project, the bigger the project team must be;

- The bigger the project team, the more important the project leader;

- Team meetings are also opportunities for orientation and training.

The proven facts, according to author Glenn Parker, are as follows [1]:

- The optimal size for a team is 4 to 6 members;

- As team size increases beyond the optimal, productivity decreases;

- As team size increases beyond the optimal, accountability decreases;

- As team size increases beyond the optimal, participation and trust decreases.

Communication among team members is a driver in establishing the optimal team size. As the number of team members increases, the number of communication paths among team members increases exponentially, not linearly. And this only addresses communication within the team. If intrateam communication becomes increasingly complicated, how much more complicated is communication between the team and the outside world?

It is a fact that a team of four to six people cannot complete every project. How then do you attack larger projects, yet still reap the benefits of optimal team size? What strategies are available to get the job done, but still develop optimally sized, high-performance teams?

- Limit the size of new teams to 4 to 6 members;

- Reorganize existing teams into teams of 4 to 6 members;

- Create core teams by bringing crucial functional representatives together;

- Assign decision-making responsibilities to core teams;

- Make use of subject-matter experts and temporary subcommittees.

We have now briefly looked at some strategies for getting the most out of project teams. Now we'll turn our attention to the subject of team types—what the types are and how they are structured, when they are used, and the advantages and disadvantages of each one.

Functional teams

The functional team structure is probably the one most familiar to the majority of us. Figure 3.1 shows that this structure would continue expanding downward until the lowest levels of the organization were reached. The nodes in the hierarchy are connected as you travel up and down the structure, but not across. This has the effect of blocking communication across functional lines. This is the typical hierarchical structure whereby front-line workers report to a first-level supervisor who reports to a manager who reports to a director and so on. An organizational chart of a company structured in functional teams would resemble a pyramid. As you move from the bottom of the organizational chart to the top, the number of people at each level decreases. Functionally structured organizations are advantageous from the standpoint that there is a clear chain of command and very little ambiguity about roles and responsibilities within the organization. This structure worked well over much of the last century, particularly in the industrial and manufacturing sectors. It made sense to structure organizations in this manner because markets were fairly stable, work processes were stable, and there were clear divisions of labor between different sectors of organizations. This structure is still useful today in organizations where there is stability and clear role definition. It is also useful in times where decisions need to be made at high levels of the organization and followed without question down through the ranks. Functional teams are typically the most comfortable types of teams to work on. They are usually composed of personnel who already work in the same area of the organization and are basically from the same peer group. Communication paths within the team and from the team up through the functional ranks are usually well defined.

Functional teams have major disadvantages, however, particularly in industries or situations where the business climate is changing rapidly and

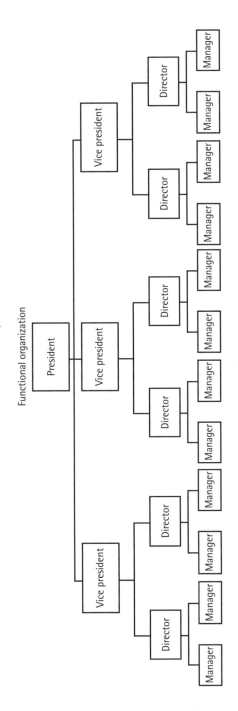

Figure 3.1   Functional team structure.

is highly competitive. Functional teams are unable to make decisions quickly at low levels of organizations. Functional teams are also inadequate for business climates that require a great deal of cooperation across functional divisions of the organization. Walls tend to develop between the functional divisions of organizations structured this way, thereby blocking the communication paths between the divisions. Communications have to travel up the chain of command in one division and then back down again in the other divisions. The process is slow and is rife with opportunities for miscommunication. Functional team organization also often leads to turf battles. Each functional division tries to act like its own independent organization, and quite often the divisions forget that they are all part of the same company. I have even seen these turf battles occur inside of functional divisions. An IT department with which I once worked was almost torn apart by the internal struggles that occurred daily between the software development group and the production operations group. The software developers resisted attempts by operations to institute operational procedures that controlled access to files and programs, while the operations staff often made changes to production processes and procedures without consulting the developers as to potential application impacts. As a result, the two groups, instead of working toward a common goal, worked against each other which impacted productivity and ultimately lead to a reduction in force of both the development and operations staffs. Their functions were transferred to other locations that had proven they could get the job done. As another example, consider Chrysler in their attempt to develop a successful small car. Their small-car division languished in obscurity and registered disappointing sales year after year. The problem with the division turned out to be the rigid hierarchical structure of the organization. The difficulty in communicating between design and production and sales, and the mistrust between those groups in the division, prevented Chrysler from being successful at producing a popular small car.

Self-directed teams

Self-directed work teams are a radical departure from the traditional hierarchical structure and allow for free-flowing communication between each team member and between team members and the client community (Figure 3.2). They require upper management to shift from a highly

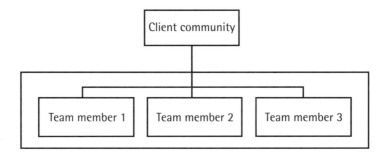

Figure 3.2 Self-directed team structure.

directive management style to a coaching and facilitating style. This is not an easy task for management in organizations where the functional team structure is firmly entrenched. Self-directed work teams also require a shift in thinking on the part of front-line workers. They must be willing to accept more responsibility for their actions and must be willing to make more decisions for themselves. The team members must also be willing to put aside personal aspirations, at least temporarily, for the good of the team. These types of teams typically are most successful in high-tech settings, or in research and development, where the individual team members are already highly motivated and work well without much direct supervision. Self-directed teams also are more successful in cultures where individualism is less pronounced, such as in Asia. These types of teams can work in Western culture, but care has to be used to insure team goals are kept above individual goals.

A real-life example of self-directed work teams should help bring this discussion into focus. An IT shop, at which I was a software developer, lost the manager of the mainframe systems programming group to a promotion at another location of the company. Rather than promoting one of the current systems programmers to the manager's position, the decision was made to reorganize the team into a self-directed work team. Of course, one of the motivations for doing this was to save the money and time that would have been spent in moving someone into the manager spot and then backfilling that person's position. However, the option of the self-directed work team would not have even been considered if it weren't for the fact that all the members of the group at that time were highly qualified professionals and highly motivated individuals who were all doing a similar type of work. The team actually pulled together nicely and worked well as a self-directed team. Each person was willing to take full responsibility for his or her actions and each person was willing to perform his or her share of the total workload. Self-directed work teams can be effective given the right group of people in the right circumstances.

Cross-functional teams

Cross-functional teams are the latest buzzword in team structure these days. Figure 3.3 provides an example of an organization divided into cross-functional teams. Depending on the organization and the industry, there might be more or fewer or different types of teams. Teams are intimately connected with the clients and with each other to form a cohesive, tight-knit organization.

The move toward this type of structure is being driven by the rapid pace at which the business climate, particularly the high-tech business climate, is changing today. Decisions have to be made faster, products and services have to be brought to market faster, and organizations have to be able to adjust their business plans faster. The rigidity of the functional structure does not allow for this quick movement and quick change of

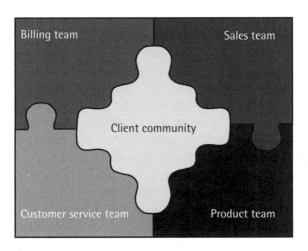

Figure 3.3 Cross-functional team structure.

direction. It also does not allow for the free flow of communication between functional areas that is critical for moving projects along more quickly. Cross-functional teams also seem to me to be the most natural fit with the way work actually occurs. Organizations don't really function by performing accounting, or marketing, or service provisioning. They really function by providing a product or service to the end customer, billing and collecting payment for that product or service, and providing ongoing customer support. Doesn't it make more sense, then, that we should organize our companies along these lines? Instead of having a marketing director and an accounting director and a construction director, why don't we have a director of billing, a director of sales, and a director of customer support. In this manner, the organization could more effectively oversee work processes end to end, rather than only managing pieces of a work process at a time.

Cross-functional teams are not without their drawbacks. First and foremost is the pain of the transition from a functional structure to a cross-functional structure. Decision-making power is transferred from the high-level functional managers down the organization to project managers or possibly front-line supervisors. This type of structure requires a different style of leadership as well. No longer are the decision-makers in charge of one functional group of people. They are in charge of a project or program team with a more diverse set of skills among their reportees. In addition to all these obstacles, communication must be managed more formally and, at the same time, communication on the cross-functional team is much more critical. It is the communication across functional lines that is the major reason for organizing cross-functionally in the first place.

Given the drawbacks to cross-functional organization, you may ask yourself if the investment in this structure is worth it. I can answer resoundingly that in the high-tech world you cannot afford to ignore this type of team organization, even if you only use it on a project-by-project basis. Refer to my example earlier in this chapter concerning the predicament Chrysler's small-car division found itself in. The solution they found to their problem of not being able to get to market quickly with a viable small-car offering was to form a cross-functional team to address the issue. The Neon was the result of this effort and it proved to be quite a success for Chrysler. I personally have been involved in numerous cross-functional teams dealing with product development in the telecommunications industry. I do not see any way these products would ever make it to market if the cross-functional approach was not used. The introduction of a new telecommunications product touches virtually every functional area in a telecomm carrier's organization. Even with the cross-functional approach, it is a major initiative to introduce a new product. The barriers that exist between areas in a functional organization would render an initiative such as this impossible.

Hybrid teams

In reality, organizations use a mixture of team types depending on the team members and the situation of the particular project (Table 3.1). Each type has its advantages and disadvantages. In some cases a hybrid mix of types is the way to accentuate the advantages of types while downplaying the disadvantages. One example of this is the use of a matrix organization. A matrix organization retains the reporting structure of the functional organization while making use of the communication and decision-making advantages of the cross-functional structure. In a matrix organization, project team participants are selected from the appropriate functional areas and set up to report to a project manager for the duration of the project, at least for their daily work assignments. The functional managers still retain reporting authority over the participants for all other aspects of their organizational lives. Typically, project managers have limited authority in a matrix organization. Functional managers still control the purse strings of the organization and they typically have the final say-so over which of their reportees actually get placed on a particular team. The functional managers also have the authority to pull personnel back off project teams if they deem it necessary to do so. Matrix organizations can be effective if functional managers buy-in to managing initiatives with project managers and sound project management principles. My personal experience with the

TABLE 3.1   Matrix Structure

	ACCOUNTING MANAGER	INFORMATION SYSTEMS MANAGER	MARKETING MANAGER
Project Manager 1	Accountant 1	Programmer 1	Sales Executive 1
Project Manager 2	Accountant 2	Programmer 2	Sales Executive 2
Project Manager 3	Accountant 3	Programmer 3	Sales Executive 3

matrix structure was less than satisfactory. The biggest drawback I found while operating under this structure was that invariably you encounter a conflict between the desires of the project manager and the desires of the functional manager and you are caught in the middle. If you please the project manager you run the risk of alienating the functional manager, which is bad for you once you come off the project and start reporting completely to the functional manager again. If you side with the functional manager, you might alienate the project manager, adversely impact the review of your project work from that project manager, and adversely impact the work of the project itself. It's the old conundrum of trying to serve two masters.

In a matrix structure, project managers are in charge of team member work assignments while functional managers are in charge all other aspects of the team members' organizational lives. This structure is a hybrid between the functional and cross-functional structures.

Another hybrid structure employed by some organizations is a combination of cross-functional project teams and functional teams. This is different from the matrix structure in that the cross-functional teams are stable, permanent teams, and the functional teams are stable and permanent as well. The functional teams provide services that seem to make more sense coming from a functional area rather than being a part of every project team. One excellent example of a type of function that might be handled this way is procurement. If most project teams require common materials and resources, it may make more sense to establish a functional procurement area to handle the needs of all project teams. Economies of scale can be recognized with this structure, fewer personnel can be employed to do the same amount of work, and specialized expertise can be developed in the functional area. Other candidates for this type of structure would include any functional area where particularly specialized knowledge is needed, but in smaller quantities than would be needed if every project team had to hire the expertise. Further examples that come to mind are the legal, security, and human resources functional areas.

Critical success factors for teams

It is critical to select the right team structure for the given project situation and for the given project team. This is the first step in building a project team with the capability to succeed. Beyond the actual structure of the project team, there are seven factors that are critical to the success of the team:

1. *Build a strong business case.* No project should be undertaken if there is not a compelling reason to perform the project in the first place. The team should be able to express the expected payback of the project in some detail. The team should also be able to verbalize the probability of achieving the payback.

2. *Gain organizational commitment to the team.* Team members must be committed to the team and the organization as a whole must be committed to the team. The right mix of personalities and skills must be put together for the task at hand. The team must be left free by the rest of the organization to make decisions. The organization also should commit to putting and keeping the right people on the team. Project team turnover can be disastrous if it gets out of hand.

3. *Give the team the confidence that it can succeed.* The team will be much more confident if they are empowered to create the project mission, the critical success factors, the project scope, the project change management process, and the project communications plan. If the team feels it has ownership of the project process, it will have the confidence to get the job done, and subsequently the project effort will be successful.

4. *Partner with the project customer.* The project team must develop a "we" mentality with the customer as opposed to an "us versus them" mentality. Customers must be involved in the decision-making process every step of the way, so as to avoid surprises at the end of the project. How many times has it happened that a project team produces exactly what the customer asked for, but not what they wanted? Customer involvement will help eliminate guesswork on the part of the project team, thus allowing the team to concentrate on executing the project plan.

5. *Partner with external vendors.* By external vendors I mean anyone external to the project team who will provide a product or service to the team in furtherance of the project goals. Just as the project team needs to keep the customer in the loop on project progress and

decisions, so too should external vendors be made to feel that they have a part in the success of the project. Demanding that external vendors deliver products and services at particular times and in a particular manners is no better than a project sponsor setting an end date and requirements before the project team has had a chance to investigate the expected output of the project process. Teams need to negotiate with their external vendors, they need to monitor the progress of their external vendors in delivering products and services, and they need to assess vendor performance beyond delivery.

6. *Enable team members to succeed.* One of the toughest, but most important, jobs of a project leader is to motivate the members of his or her team to perform at peak levels. A project leader should strive to do the following:

   • Connect project goals and individual team member goals;
   • Use facilitative leadership to solve problems and make decisions;
   • Focus on small incremental results; they help build confidence;
   • Maintain an open-door policy, and solicit feedback on what's working and what's not;
   • Show appropriate appreciation and recognition early and often.

7. *Encourage personal growth.* Motivate team members to develop a personal vision of what they want to achieve and help them integrate that into the work you're performing on the project team. In addition, as a project leader you should have a strong sense of where you want to go as well. You can't very well lead others if you don't know where you personally want to go with your life.

Reference

[1]   Parker, G. M., *Cross-Functional Teams: Working with Allies, Enemies, and Other Strangers*, San Francisco, CA: Jossey-Bass Publishers, 1994.

4

Stages of team development

Introduction

Project teams, regardless of their type or purpose, tend to exhibit a common life cycle. They experience birth, exploration, discovery, performance, and, finally, completion. Each phase of the cycle is necessary and each phase requires a different style of management. A project leader must recognize the phase the team is experiencing at any particular moment and adjust his or her style accordingly. He or she must also work to make the transition from one phase to another pass as smoothly as possible. In particular, the project manager must move from birth to performance as efficiently as possible so that the team may become productive as quickly as possible. If a team becomes mired in any early phase, or if it regresses from a more advanced phase to a previous phase, time on the project will be lost. Resources that were expended to get to that point in the project will be lost as well and additional resources, not originally planned for, will be required to move the team in the right direction again. Let's explore these stages of team development and the management styles that should be employed at each stage (Table 4.1).

TABLE 4.1   Cross-Reference of Leadership Styles with Team Stages

	LEADERSHIP STYLES				
	DIRECTING	COACHING	SUPPORTING	DELEGATING	SUPPORTING
	High directing/ Low supporting	High directing/ Low supporting	Low directing/ High supporting	Low directing/ Low supporting	Low directing/ High supporting
TEAM STAGES	Birth	Exploration	Discovery	Performance	Completion

Birth

A number of terms exist for the beginning stage of team development, such as form, orientation, creation, and kickoff. I prefer the term birth because of the connotations of new beginnings and fresh starts. A birth implies that a new, unique entity is coming into being without preconceived notions of how the world should work. The community in which this entity exists is excited at the prospect of what it may become. The community is typically eager at this point to nurture this new being and contribute to its growth and development. On occasion certain members of the community may work against the formation of this new entity. Just as a mother must follow a certain regimen before the birth of a child to insure that its birth is as complication-free as possible, so must the project leader begin to work as soon after project selection as possible to insure a complication-free project team formation. He or she must be attuned to the political climate into which this team is being introduced, and he or she must use all the risk management and conflict-mitigation strategies available to avoid the possibility that the team will get started on the wrong foot. The birth analogy works on many levels in describing the beginning stage of team formation.

Staffing the project team is a major milestone at this point in the project life cycle. In an ideal situation a project charter will be available, a total requirements statement for the project will have been developed, basic work packages for the project will have been defined, deliverables will have been identified, and a budget will have been developed. The project leader could then begin to interview potential candidates for spots on the team based on how well these candidates' abilities meshed with the project tasks to be accomplished. Of course, we don't live in an ideal world. Often requirements, deliverables, tasks, and budgets are not complete before the project team has to be assembled. Project leaders don't always have the opportunity to select their team members either. Many times people are placed on a project team because they happen to have the time to participate at that moment. A project leader should do everything within the scope of authority to influence the selection of project team members. After members are selected for, or assigned to, the team, the project leader can then move on to the next tasks of developing team cohesion and team harmony.

The members of a newly formed team are generally eager to participate at this point in the cycle. They have high, positive expectations of what the team will ultimately accomplish. At the same time, team members may be anxious about their role and about how they will fit in with the other members. The primary issues that must be dealt with at this stage are the extent to which members will place their trust in the appointed leader and their willingness to participate in decision making and problem solving. Other sources of anxiety include how the project will impact team members' professional growth, how the project will affect team members' personal life, and how the work of the team will impact team members' day-to-day work activities. Many times people are placed on project teams, but are still expected to maintain the same level of day-to-day duties as they always have. Even more vexing is the situation where a team member has to simultaneously participate on a project team and train a less experienced employee to perform their old job functions. Morale is usually high among team members despite these anxieties. Task accomplishment at this point is low with most of the energy of the team going into defining goals, tasks, and member roles.

A project team leader in the birth stage can take advantage of the high energy and morale of the team. It is imperative that the leader exerts influence at this stage and helps the team allay the anxieties that exist. The most important tasks for the leader to accomplish at this time are to define the goals of the team, to help team members discover who they are, to help team members discover who their fellow team members are, and to define the roles and responsibilities of each team member. Many instruments exist to identify the personality traits of team members. These instruments can be an invaluable aid both to the project team leader and to the team member. The leader can use them to learn how best to interact with the team members and what tasks each member may be best at performing. The team members can use the results of the instruments to learn more about each other and how best to facilitate communication between each other. Examples of these instruments include the following:

- *Myers-Briggs Personality Assessment* [1]. This instrument assigns a personality style based on the participant's answers to a number of multiple choice questions. According to this survey, individuals fall into one of sixteen different personality styles. The styles reflect whether an individual is introverted or extroverted, sensing or intuitive, thinking or feeling, and judging or perceiving. This instrument is particularly useful as an aid in learning how to communicate with other team members based on the characteristics of their personality type.

- *Fundamental Interpersonal Relations Orientation—Behavior (FIRO-B) Report* [2]. FIRO-B reports a measure of a person's expressed inclusion, control, and affection and their wanted inclusion, control, and affection. It measures how you tend to behave with others and what you seek from them. It helps increase team members' understanding of interpersonal issues between themselves and can be an aid in conflict resolution.

- *Parker Team-Player Survey* [3]. This survey is an assessment of an individual's strengths as a team player. It categorizes people as contributors, collaborators, communicators, or challengers. The results can be used to identify what role individuals will be best at playing as members of a team.

- *Strength Deployment Inventory (SDI)* [4]. The SDI is an assessment of the strengths you use in relating to others, both at times when things are going well and at times when there is conflict and opposition. The inventory categorizes individuals into color categories (red, green, blue, and all the combinations of these colors) to represent the dominant elements of the team members' personalities under given conditions. This instrument is invaluable in understanding why certain people react the way they do in certain situations. It also gives you insight into how to deal with individuals based on the given situation.

Once these instruments have been administered and the results reviewed, the team lead and the team members should have a much clearer picture of their fellow team members and how to interact with them. Team members' anxiety in this regard should be greatly eased and the project team leader will have an easier time moving on to goal development and role assignment.

The other major hurdle to overcome in this stage is that of team and project goal setting. Clear, concise targets should be set for the team at this time. It is much easier to concentrate your efforts if you have a clear understanding of where it is that you are trying to go. Teams are doomed to flounder aimlessly without a strong statement of goals and the ultimate mission of the team. This fact has been proven time and again in the business world, the sports world, and in the military. The SMART acronym should guide goal development. Goals should be specific, measurable, achievable, realistic, and time-bound. Normally, a kernel of a mission statement and list of goals is provided by the organization sponsoring the project, but it is up to the project team to flesh out these elements. The project leader must exhibit a highly directive management style at this stage in order to quickly and efficiently establish the mission and direction of the team. Sample mission statements include the following:

- To be a world-class telecommunications company, the standard by which others are measured.

- To be the number one aerospace company in the world, and among the premier industrial firms as measured by quality, profitability, and growth.

- To provide timely and innovative business solutions that exceed the needs of the marketplace, while seeking enterprise-wide collaboration, and creating exceptional satisfaction for our employees and customers.

A final element to consider in the birth stage is that of a communication plan. For example, how often will the team meet? Where will the members meet and what will be the format for the meetings? Should team members communicate strictly in writing or will a certain amount of informal, verbal communication be permitted? How often will sponsors, stakeholders, and supporters be updated on project progress? What form will these updates take? Team members must know how to communicate with each other and with influences outside the team. Sponsors, stakeholders, and supporters will also want to know what sorts of communication to expect from the team and how they will receive communication. Sometimes these external groups will define their communication needs and the team will have to plan for how to meet these demands. The team will also want to make sure they have a vehicle with which to communicate their successes and through which they can alert external contributors to potential problems. An adequate communication plan helps set expectations for all parties involved in a project and is one item that can be addressed and resolved early in the project before the rigors of actual project performance begin.

Exploration

The second phase of the project team life cycle is sometimes referred to as dissatisfaction or storming, but I prefer to use the term exploration. This stage only degenerates into dissatisfaction if the project team members are not shepherded through correctly by the team lead. It is at this point in the project that the initial idealistic expectations meet with a cold, hard dose of reality. The kick-off meetings are over and the time for real work has arrived. Team members need time to explore the landscape and try to find their niche in the team. The results of the survey instruments from the birth phase should be a tremendous aid in this exploration. The team members can begin to try approaches to communication with one another and the team lead can experiment with conflict resolution strategies. The issues of power, control, and influence will be foremost in the team's collective mind at this stage. The team lead's authority may be challenged and it is imperative that the leader exercise his or her power with a heavy dose of common sense thrown in so that the team members might be reassured that someone is managing the direction of the team. Team members will be confused and frustrated at the underdeveloped structure of the team at this point and discouraged at the long road between them and project completion.

The exploration phase in some ways is the most critical phase in a project team's development. Studies in the early 1990s by the Department of Defense over a 13-year period show that you can estimate with considerable accuracy whether a project will meet its completion date after only 15% of the project is complete. After only 20% of the project is complete the cost performance efficiency factors have been shown to be stable. If you are behind schedule at the 15% mark, chances are you will not make your deadline. Exploration is that phase of project development and project team development that determines whether you are on track at the 15% line. The newly developed communication plan will get its first test at the exploration stage. Parties both within and without the team will attempt to circumvent the formal communication paths believing that they are just getting the job done in a more efficient manner. The opposite effect will be realized, however. All team members will not know the same information at the same time and resources will be needlessly expended bringing everyone to a common understanding of the current project status. Team members may begin to form subgroups within the team based on allegiance to a particular vision of how the project should proceed. These subgroups may also form around particular personalities within the group, a trend that can become very counterproductive over the life of the team.

It is apparent now that the project team leader must firmly grasp the reins of the team from the outset, or he or she will soon lose control of the group. A number of strategies are at the team lead's disposal to aid in moving the team successfully through the exploration phase:

- *Reaffirm the team vision, goals, expectations, roles, and relationships.* The team is searching for footholds in the exploration stage. A strong, directive management style is necessary to help team members know with certainty where they are going and how they will get there. One way to keep the vision and goals in the forefront of the team members' minds is to repost them at every team meeting. One major telecommunications company posts its corporate vision, mission, and strategies in every conference room in every building in the corporation.

- *Encourage and support interdependence.* Team member personalities have been identified and are beginning to manifest themselves by this time. The team lead should try to get the best from this situation by encouraging the members to rely on their survey results and put them to use in learning to deal with each other and learning to mesh as a team.

- *Practice active listening.* The team lead should treat team member suggestions and complaints uniquely and individually. Active listening builds cohesiveness between the leader and the members. Part of the communications plan should address how team members should approach the team lead with suggestions and concerns. The old axiom, You have two ears and one mouth, so you should listen twice as much as you talk, applies here.

- *Provide skill development and decision support.* In addition to a strong, directive leadership style, a team lead in the exploration phase needs to employ a coaching style at the appropriate time also. Team members need to feel that they will get all the training and support they will need to perform their roles on the team, and they need to know that the team lead will support them in the decisions that they make. I had the opportunity to participate on a team whose job it was to transition some regionally developed data processing applications into an existing corporate system. The team leads at corporate headquarters made a number of assumptions about the level of understanding the regional team had in dealing with corporate data processing procedures. Unfortunately, they overestimated our knowledge of the corporate procedures and so we spent a good bit of effort spinning our wheels in an attempt to perform the tasks we had been given. Our local team lead attempted to obtain the training we needed, but was met with indifference at the corporate level. The local team members went away from this project at its conclusion confused and bewildered, and in no better position to function on a similar type project than they were when this project began. Don't let your teams founder like this. Make sure early on that training requirements are identified and then follow through to provide that training. Time spent on training at this point in the game will pay off in efficiency in later stages. I was fortunate enough to experience the benefits of this proactive approach on another project to which I was assigned. Technical and functional members of the organization spent a week with my team reviewing, in intricate detail, all aspects of the system for which we were about to assume support. The morale of the team was sustained by this show of support and we were able to become productive within weeks of assuming support of this particular system.

- *Praise constructive behavior.* Start the reward system for your project early. Take time to recognize good ideas and team play from the beginning of the project. Team members will perform even harder

later in the project if they have proof that their hard work will be recognized and appreciated. Find ways to reward and encourage team
behavior while recognizing individuals who stand out above the
crowd. Be creative in your reward structure and make rewards
meaningful. A day off with pay may be more meaningful than a
dozen plaques presented by company executives. Rewards can be as
simple as a thank-you note from the project lead, or as complex as an
all-expenses-paid vacation. Sometimes a team can be rewarded for a
job well done, and individual performance can be rewarded with
extra perks at the same time. An acquaintance of mine once managed a large computer installation project for the United States
Navy. One of the requirements of project implementation was for
team members to be onsite at various locations around the world
during the installation. All the team members reaped the benefits of
traveling to various naval installations around the world, but the
individuals whose performance stood out were assigned to such
locations as Italy and Hawaii.

♦ *Manage conflict.* Conflicts will develop. Be prepared to deal with
them before they start. Recognize that there may be more than one
way to solve a problem. Two individuals can have a difference of
opinion and both individuals may be right. Every issue is not necessarily black and white. Recognize the shades of gray. Confront difficulties head on when they first appear. Pots moved to the back
burner for too long often boil over and demand attention, whether
you have the time to deal with them or not. I recall an experience
some years ago where one of my peers was on the verge of resigning
because she did not feel she was being given challenging assignments, yet our team had more work to accomplish than we had
resources to whom to assign the work. Our project lead was not
passing assignments to this individual because he was under the
impression that she was dedicated 100% to another project, when in
fact the other project was consuming very little of my peer's time.
Had the project lead taken the time to deal with this situation early
on, he could have avoided an uncomfortable period of conflict with
my peer. Instead, he chose to ignore the chasm that was developing
between himself and his subordinate, until the situation came to a
head and he was forced into action at a time when he already had
more on his plate than he could handle.

The ultimate goal of a project team leader in the exploration phase is to move through the phase as quickly and as painlessly as possible. The team lead will be called on to exercise every ounce of human resource skill at his or her disposal. Be assertive and give the team a leader to follow, while at the same time make sure not to leave anyone behind. Manage the project schedule carefully through this phase and keep your team members apprised of any schedule or budget slippage. The team should be ready to perform much meaningful work as it emerges from the exploration phase. Union Carbide Corporation uses a team development process that encompasses much of what we have discussed here. The Team Development Process helps form a group into an aligned, focused, and motivated work team that strives for a common mission and is capable of delivering improved project results. The Team Development Process has three primary elements, each of which contributes significantly to the success of the process as a whole. The first element focuses on integration with the project work process. It is a work-oriented planning component that defines where to go, how to get there, and how to recognize arrival. It is project-specific. Each project adapts the project work process to its particular needs and constraints. It focuses on the resolution of real issues and problems facing the team. The second element focuses on team dynamics and human interaction. It includes application of team skills such as group decision-making and problem solving. It stresses the importance of active listening and effective meeting management. It examines how communication styles affect ability to relate to each other and to resolve conflicts. An important focus of this element is role clarification and expectations exchange between functions. The third element focuses on performance measurement and recognition. It provides an opportunity to recognize the entire team for performance in key result areas against set measures. It also provides an opportunity to recognize individuals and teams for implemented suggestions that improve project performance in predetermined areas. [5]

Discovery

The discovery phase of team development can be likened to watching the first rays of daylight appear from beneath the horizon. Responsibilities have been assigned in the exploration phase, political conflicts have been ironed out, and an initial road map for the project has been defined. Answers to the exploration phase questions have now been discovered, hence the name of this phase. Team members exhibit less dissatisfaction with the current state of the project. The expectations of the project and the realities of time and budget constraints have been brought closer together. Team member skills have been aligned with project tasks so that project goals appear to be attainable. Team members' morale and self-esteem is on the rise and the members are more confident that they can perform their assigned work.

The personality clashes common in exploration have now subsided, as team members are more willing to reveal their inner selves. There is a higher level of comfort between group members as the team learns to function as a team and not a collection of individuals. Task accomplishment ramps up as the meaty work of the project gets underway. Everyone is familiar and comfortable with the communications plan established at birth and information flows more easily both inside the team and between the team and the outside environment.

The team leader shifts his or her attention in the discovery phase to concentrate more on project task accomplishment and less on team building. The leader cannot ignore the needs of the team however. He or she still must practice good active-listening skills and he or she still must provide encouragement and support to team members. While the team is on the verge of high performance, it is not there yet, and the team lead must continue to provide the leadership necessary to keep the team moving in the right direction. The team lead should involve team members in the decision-making and problem-solving process and should continue to reward the efforts of the team members. Positive reinforcement of good work ethics and good team skills is a powerful tool at this stage of team development. Allow team members to participate actively in progress reports and update meetings with stakeholders. Avoid the trap of isolating the team from the outside world. Team members need to be given enough insulation to perform their assigned tasks, but they also need the opportunity to hear directly from sponsors, stakeholders, and supporters. Ideas, suggestions, and complaints often carry more importance when team members can hear them directly from the outside world rather than through the filter of the team lead. Team members also need the opportunity to toot their own horns from time to time. Team members who have solved a particularly difficult problem or who have put in extra effort to keep a project on track will derive a great deal of satisfaction if they are allowed to report their successes directly to the outside participants. This will also avoid the possibility that the team lead will be given, or worst case take, the credit for team successes.

Performance

The fourth stage of the team development cycle is performance. Team members continue to be confident about their roles and their relations to others on the team. Morale continues to be high and the project becomes self-rewarding to a degree because milestones are now being passed. Team members are sure of their abilities and contribute freely to the work of the project. They work more autonomously and do not depend as much on the team lead for support. Members now begin to recognize each other's accomplishments and also begin to challenge each other to achieve even more. Communication within the team flows freely, but the actual amount of communication may decrease slightly because team members can anticipate each other's moves now. The energy of the team is at its peak as the goal of the project is now in sight.

The project team leader in the performance stage must employ a delegating approach to team management. The team is at a high-performance levels and the team lead must know when to leave well-enough alone. High performers will resent a leader that attempts to use a highly directive style of management. The very thing that makes a team member a high performer is that he or she already knows what needs to be done and how to do it. It is a waste of resources for the team lead to detail every task for a high-performing team member. The leader can build on the energy and the morale of the team by delegating tasks and relying on team members to perform work with little or no supervision. The team will recognize rewards for their work when the leader delegates authority to them. The leader should concern himself in this phase with monitoring project status and making sure no unforeseen risks crop up that might jeopardize the end date of the project. Status updates to project stakeholders continue during this phase, and team members should be allowed to participate in these meetings. The team lead should keep a keen ear tuned to the stress level of the group during the performance phase. It is common in this phase for team members to begin experiencing some burnout from their assigned project tasks. Nerves can become frayed if a large amount of tasks have to be accomplished in a short amount of time. The team lead should resist the urge to immediately jump in and take over in these situations. Instead, he or she should help the team members to momentarily step back, catch their collective breath, and then to move forward again. Sometimes a small break in the action can prevent a larger breakdown closer to the end of the project. The team lead should also be making preparations for implementation of the project while the team members are finishing their project tasks. An implementation and production turnover plan needs to be in place prior to project completion. The leader should make sure both the team members and any involved outside parties are apprised of the implementation far enough ahead of time to allow for a smooth transition from project to production.

Completion

Once a project's goals have been met and the product or service the team worked to develop has been implemented, the project team will pass through the completion phase of its life cycle. This is the phase that most often is forgotten or ignored, but it is an important phase and the success with which you handle completion can impact future projects for your organization. Team members will experience both strong positive and strong negative feelings during this phase. They will be happy to have achieved their project goals and should take pride in the contributions they have made to the organization. On the other hand, they will be saddened by the impending separation of the team and may feel apprehension over what lies beyond the bounds of the current project. Morale and energy level begin to decrease during completion and productivity falls away.

The team leader must manage the completion phase just as carefully as the other four phases. He or she must make sure that the team stays together and maintains its focus long enough to completely finish implementation. A postmortem should be conducted soon after project implementation so those lessons learned by the team members may be documented for future project team reference. Overall project success should be rewarded as soon as possible after project implementation. Many times team members disperse and move on to other projects before the rewards for previous projects are passed out. Organizations lose the benefit of the positive reinforcement of the reward when the action and the reward are separated by long stretches of time. The project leader should also work quickly to make sure project documentation is completed during the completion phase before team members disperse. This documentation should have been compiled over the course of the project, but it must be finalized and officially published during completion. Future project teams may save precious time and resources if they have access to adequate project documentation from previous projects. Finally, the team lead should take time to produce a thorough performance review on each team member. This may be a stand-alone review or it may contribute to a comprehensive yearly review for the team members. Team members will receive more accurate and more timely reviews if the project team leader prepares the review during completion, rather than waiting until a later date.

Conclusion

The basic life cycle of a project team is predictable, but the activities of any individual team are not. Project team leaders must understand the different phases of the cycle and must be able to identify which phase their team is in at a particular time. Distinct management styles are required for each phase and the use of the wrong style for a particular phase can spell disaster for a project. The project leader must also understand the personalities of the individual members of his or her team and use that information to aid him or her in communicating with the team members. The team lead must understand when the team needs him or her to assert his or her leadership and when to back off. He or she must understand when a team member needs coaching and when the member simply needs support. Lastly, the leader must orchestrate a successful completion to the project so that team members go away fulfilled and stakeholders are left with the desired project outcome.

References

[1]   Myers, I. B., *Myers-Briggs Type Indicator*, Palo Alto, CA: Consulting Psychologists Press, Inc., 1985.

[2]   Schultz, W., *FIRO-B: Self Scorable Booklet and Answer Sheet*, Palo Alto, CA: Consulting Psychologists Press, Inc., 1996.

[3]   Parker, G. M., *Parker Team Player Survey*, Tuxedo, NY: Xicom, 1993.

[4]   Porter, E. H., *Strength Deployment Inventory: Manual of Administration and Interpretation*, Pacific Palisades, CA: Personal Strengths Publishing, 1992.

[5]   Piper, R., *The Triad and the Evolution of Project Management*, Newton Square, PA: Project Management Institute, 1996.

5

Communication systems for teams in a technical environment

Introduction

So far we have established working definitions for projects and teams. We have examined different types of teams and the stages of team development. The next few chapters will look at a series of topics related to project teams as they perform their project tasks. We'll look first at communications. Anyone who has ever served in the military can tell you without a doubt that the key to success in battle is communication. The high-tech business world is no different. Communication is an essential element of success in the fast-paced environment in which technical project teams operate these days. The best laid plans of a project manager and his or her team will surely go awry if the communication systems employed by the team are not effective.

Much of what we learn about communicating is learned before we are two years old. We have been practicing and refining these skills longer than just about any other human skill we utilize. It stands to reason, then, that our communication skills are some of the hardest to change. Communication is practically instinctive for most of us and we don't really examine the way we communicate. If we communicate poorly, however, or if we employ a communication style that does not fit the particular situation in which we find ourselves, we must examine our method of communication and look for improvement.

Formal systems

Before we can attempt to improve the way we communicate we must look at all the options of communication systems that are available to us. We'll start with the formal systems and then discuss more informal options.

There are a myriad of formal communication systems, and the choice you make of which system to use depends on a number of variables. These variables could include such items as personality of the person with whom you are trying to communicate, distance between you and that person, how familiar you are with that person, the type of project to which you are currently assigned, and the resources made available to you by your organization. Some of the formal systems that may be available to you are as follows:

1. *Formal face-to-face meetings.* The most natural, and probably most widely utilized form of formal communications in the business world, is the traditional face-to-face meeting. These meetings may be between only two people, between one person and a group of people, or between two groups of people. Regardless of the configuration of the meeting, someone has to take charge of organizing the gathering, and someone has to be in charge of conducting the meeting itself. I have participated in face-to-face meetings in the past where the person in charge did not really take a firm grasp of the reins of the meeting and I can assure you nothing good ever came of those meetings. In order to be successful, formal face-to-face meetings need several elements of organization:

    • *Purpose, agenda, and limits.* Meetings must have a stated purpose. Just as a project needs a mission, so, too, must meetings have a mission. The purpose should be stated clearly and concisely and should be announced to all meeting participants ahead of the meeting time. Everyone should be focused on a clear purpose when they assemble for a meeting. Face-to-face meetings should also have a stated agenda and each agenda item should have a time limit assigned to it. The agenda provides a framework to guide the discussion during the meeting and it also helps define deliverables that are expected to be produced during the meeting. The agenda-item limits help keep the discussion focused and give the meeting participants a sense of urgency about accomplishing what needs to be accomplished. The most effective meetings I have ever attended were run by a facilitator who actually timed agenda items with a stopwatch. Once the limit for the agenda item was reached, the facilitator stopped the discussion. She gave the participants the option of tabling the issue, or taking five more minutes to resolve it. These meetings rarely ran over time and they always left you feeling that you had produced concrete results (see Box 5.1).

Box 5.1   Purpose, Agenda, Limits (PAL)

Accounts Payable transition-plan update meeting
Room 3103
dd/mm/yyyy
Purpose: The purpose of the meeting is to update the Accounts Payable production-support transition plan with information current as of dd/mm/yyyy.

Agenda:

Welcome	1 minute
Roll call	2 minutes
Review open transition items	30 minutes
Update plan with new items	10 minutes
Questions and answers	10 minutes
Closing	2 minutes
Limits	

The Accounts Payable transition-plan update meeting will be held in Room 3103 and is scheduled for no more than 60 minutes. Individual agenda item limits are noted above. The dial-in number for participants not in the headquarters building is 999-999-9999, pin number 999999.

- *Facilitator.* Every meeting needs a facilitator or there is no purpose in meeting at all. Someone must monitor the agenda, monitor the limits, and make sure the discussion stays on track. A meeting without a facilitator is like a ship without a rudder. Ideally, the meeting facilitator should be a disinterested third party who is trained and skilled in the methods of quality facilitation. This allows the facilitator to concentrate on conducting the meeting, rather than having to participate in the discussion as well. It also allows the facilitator to impose some discipline on the meeting without any introduction of bias. The facilitator must be someone who can manage the variety of personalities that might be found in a meeting. The facilitator must also be someone who can command the requisite amount of authority from the other meeting participants.

- *Recorder.* An often-overlooked element of a formal meeting is the recorder. It is always a smart idea to produce minutes from any formal meeting that you conduct. Minutes should be distributed to all meeting participants as soon as possible after the conclusion of the meeting so that a written record may be established of the proceedings of the meeting and also so that all meeting participants may have a common understanding of the proceedings of the meeting. It is difficult at best to be both the facilitator and recorder and it is practically impossible to be a participant, a facilitator, and a recorder. A designated recorder can concentrate on capturing all the happenings during a meeting without regard for the context of the happenings. Once again, ideally the recorder would be a disinterested third party to the proceedings, but it is possible to rotate the recorder position among the meeting participants from meeting to meeting and be successful. The main point to remember is that every formal meeting needs someone designated as a recorder.

- *Adequate facilities.* It may seem that I'm stating the obvious, but it is essential that you have adequate facilities for any meeting that you conduct. I have attended meetings in cafeterias, in cubicles, even in hallways because the person who called the meeting in the first place did not take time to plan far enough ahead to arrange for adequate facilities. Needless to say, these meetings were less than productive. If you are planning a meeting, you need to insure that the location at which you will meet has adequate lighting, adequate space for the number of participants, and adequate seating for the participants. The site should also be in as neutral a location as possible so as not to give the appearance that one party to the meeting is favored over all the others. Also make sure you have adequate supplies available to capture the proceedings of the meeting. You only have a set amount of time in which to gather all your participants and it is a waste of their time and your organization's resources to spend part of the meeting time trying to gather what you need to conduct the meeting.

2. *Video conferencing.* Video conferencing is similar to face-to-face meetings in that the participants actually get to see each other, but you have one or more groups that are remote from each other. The participants are linked via video conferencing equipment. The same needs exist for video conferences as exist for face-to-face meetings, but their importance takes on even greater meaning. Once you put participants at a distance from each other, it is even more important to have a strong facilitator. It is even more important to have an agenda with a stated purpose and to have limits to which all participants adhere. My video conferencing experiences have been much less successful than my experiences with face-to-face meetings, but the reduced effectiveness of video conferences are usually offset by the time and money savings realized by not pulling all your personnel into one location for a meeting. The number of remote-site participants in a meeting, their distance from the meeting site, and the length of the meeting all play into the decision as to whether to meet face-to-face or participate via video conference. Despite the cost savings realized by employing video conferencing, many organizations still spend huge amounts of money on travel and expenses because they refuse to make full use of the medium. Many people are uncomfortable in front of a camera and so they are reluctant to participate in video conferences initially. Organizations should work to alleviate these feelings in their employees so that they can take advantage of the enormous resource savings afforded by video conferencing.

3. *Teleconferencing.* A third form of formal communications is teleconferencing. Again, this is a step down from face-to-face meetings, but it is also much less expensive than video conferencing, and sometimes it is less expensive than face-to-face meetings. A teleconference is actually a very good vehicle for certain types of formal communications. I am currently part of a team of people supporting a suite of applications. Each day this team utilizes a teleconference to review the status of the previous night's application processing. The call rarely lasts more than ten minutes and none of the participants have to leave their desks. Teleconferences are ideally suited to this type of short, routine meeting. Generally a teleconference works well when your meeting topic or topics are well defined, the limits for each item are short time-wise, and the number of participants is limited. It is essential that you have a well-defined agenda for the teleconference and that you have a facilitator who can guide the call. Participants must learn to identify themselves when they speak and they must learn to speak in turn. Sometimes teleconferences will be conducted as a way to link groups of people as opposed to individuals. In these situations the physical facility for the meeting becomes an issue. Groups should be kept small and the meeting area must be arranged such that all participants can sit equidistant from the telephone. It is a waste of time for a participant to sit so far away from the telephone that no one else on the call can hear what he or she has to say. Not only that, when participants on a call are not speaking loudly or plainly enough, invariably sidebar conversations crop up and at this point the facilitator has lost control of the teleconference altogether.

4. *On-line chat sessions.* A final form of formal project team communications I would like to discuss is a relatively new phenomenon. The explosion of connectivity and use of the Internet has given rise to the development and use of chat rooms. These are designated areas in cyberspace where participants can log in concurrently, type messages on their computer, and transmit those messages for all participants to see on the chat screen simultaneously. Chat sessions are similar to teleconferences in that they work well for short-duration, well-focused discussions. The agenda should be set well ahead of the chat session and should be distributed to all

participants. A facilitator is an absolute must for a chat session to be successful, otherwise no one will know when or where to begin the discussion. Ideally the facilitator should begin by welcoming all the participants, then concisely stating the purpose of the discussion. The facilitator should then address each participant in turn for input to the discussion. Limits should still be imposed on agenda items, but the facilitator should take into account the fact that participants are having to type in all their responses and should adjust the limits accordingly. Chat sessions take some getting used to. Many people do not like them because you are completely removed from the human intervention factor that is present with the other forms of formal communications discussed so far. It takes some practice to know when to talk and when to listen. It also requires a certain level of comfort with on-line interaction and it helps if you are a competent typist. I have firsthand experience with chat sessions and can attest to their success given the willingness of the participants to make them work. My experience was with four other people who were engaged in a training class with me. Our assignments for the class required us to meet on a weekly basis to discuss the work to be completed that week and to come to a consensus on the position the team would take in answering the challenge of the weekly assignment. It only took one session for us to discover the necessity of a well-defined agenda and a facilitator. We rotated the facilitator position among the group members over the course of the class, thus providing everyone with the opportunity to be a participant and a leader. Our group was able to develop solutions to the weekly assignments via these chat sessions and our hard work was rewarded with high grades from the instructor at the end of the course.

5. *Reports, charts, and graphs.* Certainly one of the most important formal, nonverbal forms of project communication is graphical output from the project management (PM) software utilized on the project. Modern project management software allows for a multitude of reports, charts, and graphs to be generated. A workbreakdown structure (WBS) is the beginning data loaded into the software, and from there resources are applied to the tasks on the WBS. Gantt charts, Pareto charts, network diagrams, and resource

usage charts are just some of the graphs and reports modern PM software can provide. These graphical outputs can be used to show project progress, critical project tasks, and resource utilization to project stakeholders who in turn may wish to make business decisions based on the information provided. Tables 5.1 and 5.2 and Appendix B provide information about specific softwares.

Informal systems

Informal communication systems are more difficult to list because there are so many of them, and so many varieties of the myriad choices. This is not to say that informal communication systems are less important than formal systems. In fact, the informal systems may be more important. Often it is via these informal systems that important agreements are reached, breakthrough ideas are identified, and important bonds are formed between project team members and between the team and the outside world. I will touch on a few of the more common systems in use in the high-tech project world today, but remember this is only a sample.

- *Spontaneous conversation.* This is probably the most widely used form of informal communication. Sometimes it may take the form of a conversation over a cube wall, other times it may be the proverbial conversation around the water cooler. No one person can know everything that has to be known to execute a particular project, but if all the members of a project team have the capability to freely converse with each other, then each person has access to the pooled knowledge of the group. This pool of knowledge is very powerful if everyone learns to utilize it. It is very convenient to be able to just toss a question out into thin air and have a reply come back from whoever happens to know the answer. It's almost as if you had the answer yourself. Other times small groups of project members may gather in a hall or an aisle to tackle an issue. I have had many occasions where I was discussing an issue with a colleague, only to have another group member stop by to offer his or her opinion and it turns out he or she has resolved a similar issue previously. Many old-school managers frown on these informal gatherings of employees. They think that if everyone is not at his or her desk, nose down in a pile of work, then no work is getting done. I would argue that

TABLE 5.1   Satisfaction with Project Management Software

SOFTWARE	OVERALL	CONTENT	ACCURACY	FORMAT	EASE OF USE	TIMELINESS
MICROSOFT PROJECT	3.3	3.1	3.4	3.4	3.3	3.4
MICROSOFT EXCEL	3.7	3.3	4.2	3.8	3.7	3.7
ABT PROJECT WORKBENCH	3.5	3.4	3.7	3.5	3.2	3.6
PRIMAVERA PROJECT PLANNER	3.6	3.8	3.9	3.8	2.8	3.5
PRIMAVERA SURETRAK	3.7	3.6	3.9	3.8	3.6	3.9
SCITOR PROJECT SCHEDULER 7	3.8	3.8	4.0	3.8	3.3	3.9
MICROSOFT WORD	3.6	3.2	3.2	4.2	4.1	3.9
LBMX PROCESS ENGINEER	3.5	3.7	3.7	3.5	2.8	3.8
MICROSOFT ACCESS	4.1	3.7	4.9	4.9	3.5	4.1
VISIO	3.3	3.4	3.3	3.4	3.3	3.1

Source: PM Network, March 2000, p. 70.

Rankings are based on a survey of project managers, with a score of 5 indicating excellent performance, and 1 indicating poor performance.

not only should this spontaneity be allowed, it should be fostered. Project team members should be encouraged to draw from one another for both new ideas and solutions to old problems. Once a flow of knowledge is established among a group of people, it tends to

TABLE 5.2 Support for Project Management Functions

PM Function	Microsoft Project	Microsoft Excel	ABT Project Workbench	Primavera Project Planner	Primavera SureTrak	Scitor Project Scheduler 7	Microsoft Word	LBMX Process Engineer	Microsoft Access	Visio
Defining a statement of work	2.3	2.2	2.3	2.3	2.6	2.3	4.8	4.0	2.5	2.5
Monitoring change control	2.8	2.8	2.8	3.5	3.0	2.7	3.2	2.3	3.0	2.5
Decision-making	3.0	2.8	3.2	3.5	2.9	3.7	2.3	3.3	3.0	2.5
Planning and replanning	3.7	2.8	3.9	3.9	3.9	4.0	2.8	3.7	2.7	2.5
Work definition	3.0	2.4	3.7	3.7	3.3	3.4	4.4	4.3	2.3	2.8
Organization	3.2	2.6	3.2	3.7	4.1	3.8	3.0	3.3	3.0	4.3
Estimating	2.9	3.1	3.1	3.5	3.4	3.7	2.3	3.3	3.3	1.8
Recording of project data	3.3	3.6	3.4	3.8	3.8	3.8	4.4	2.7	4.0	3.0
Monitoring	3.5	2.9	3.7	4.0	3.9	3.7	2.2	2.7	4.3	2.0
Performance measurement	3.0	2.8	3.5	4.0	3.9	4.0	2.2	2.3	2.7	1.8
Problem identification	2.9	2.6	3.2	3.7	3.1	3.7	2.4	2.0	2.7	2.3
Reporting	3.4	3.4	3.5	4.2	3.5	3.7	4.4	2.3	4.5	3.3

Source: PM Network, March 2000, p. 71.

feed on itself and grow. Eventually the whole becomes more than the sum of the parts.

◆ *Ad hoc telephone calls.* Similar in nature to spontaneous conversation are ad hoc telephone calls. Often times project members are not located close enough together to allow for spontaneous conversation. Ad hoc telephone calls are the next best way for project members to stay connected with one another. It is a simple matter to pick up the phone and connect with a teammate who may be 50 feet or 5,000 miles away from you. This concept should also be pushed out to the client community outside the project team as well. I have recently been assigned to an applications team that supports a client group halfway across the country. Prior to our group taking over support, the IT support staff for the clients had been colocated with them. The client group was very apprehensive about their support being moved so far away. In initial kick-off meetings I tried to reassure the clients by reminding them that our team was only a phone call away. This did not go very far in assuaging their doubts about us. They were used to employing spontaneous, face-to-face conversations with the IT staff, and now we were asking them to employ, what was to them, a completely new method of communication. It was not until we were a couple of months into the arrangement that they realized that ad hoc telephone calls could work just as well as spontaneous conversation. Once the client group became comfortable picking up the phone, and once they realized that we would be there to answer their questions and assist them with any IT support issues, they became much more comfortable with the arrangement and with our ability to support them.

◆ *E-mail.* Electronic messaging is probably the fastest growing form of informal communications in the working world today. E-mail allows almost instantaneous distribution of thoughts and ideas from one person to another, or from one person to a whole organization. It is a convenient way to share documents and images as well. It is not as well suited to true dialogue like spontaneous conversation or ad hoc telephone calls, but it is possible to conduct dialogue via e-mail. E-mail also makes it possible to communicate thoughts from practically anywhere in the world, given that you have the ability to connect to a telecommunications network. With the advances in wireless technology there are very few places in the world today where a person cannot connect to a telecommunications network. E-mail is also a less intrusive form of communication than either spontaneous conversation or ad hoc telephone calls. The recipient of an e-mail message can choose when or if he or she will read the message and respond. E-mail offers an audit trail for both the sender and receiver. As opposed to the spoken word, the written word can be preserved for future reference. The meaning of the message may still be up for interpretation, but the content is not. E-mail can be a dangerous tool if used improperly. The recipient of an e-mail only sees the words that were written, not the tone in which those words might have been spoken. It is extremely important that you choose your words carefully when communicating via e-mail.

Internal team communication

So far we have discussed formal and informal communication systems which may be used to overcome physical barriers to communication. Other, less tangible barriers to communication also exist, which the successful project team must learn to break through as well.

- *Intellectual barriers.* Project team members must be clear as to what information is needed by whom and when. Team members must be able to constantly keep each other up to date. A neural network should be formed among the team members so that communication is flowing without the team members even having to think about it. This type of communication is not innate; it must be learned. Project managers must see that their team members receive adequate training and instruction in forming this internal communication network. The project manager must also see to it that this internal communication is monitored and feedback loops exist so that he or she is alerted if the communication system is not functioning properly.

- *Psychological barriers.* Communication barriers that arise out of personality or emotional conflicts are at best difficult to overcome. The threat posed by these barriers, whether conscious or unconsciously imposed by team members, is real and will completely disrupt the flow of team communications if not dealt with promptly and efficiently. Project managers must take it upon themselves to educate themselves on intrateam personality interactions so that they can effectively deal with these psychological barriers. The project manager must also make a concerted effort to break down these barriers between him- or herself and the project team. The project manager is just as essential a cog in the team communications machine as the team members themselves are.

* *Political barriers.* There is no escaping politics in the project environment. The political jockeying begins when the project is first proposed and continues throughout the life of the project. Project deliverables are often oversold initially in order to win project approval and so the project manager must deal with balancing what someone else may have promised with the reality of what can be delivered in the stated time frame of the project. As the project progresses, more people with various agendas will be brought into the mix, and these expectations must be melded and managed with the work in progress. Politics exist within the project team as well. Project team members may attempt to make themselves look good at the expense of their teammates. Cliques may form inside the project team that may lead to dissent and a breakdown in internal team communication. The project manager must keep his or her eye constantly on these situations and nip them in the bud before they flower out and create serious problems. Successful team building at project initiation, efficient utilization of team member resources, and liberal use of rewards across the team for accomplishments should help alleviate the possibility of political infighting among team members.

Communication with the outside world

Many of the same barriers that exist to communication within the team also exist between the team and the outside world. The project team must be clear on what information stakeholders, sponsors, and supporters need and require. A successful project plan will include a communications plan which should detail exactly what information will be provided by the team, to whom this information will be provided, when the information will be provided, and in what format. The key to a successful communications plan is to spend adequate time at the beginning of the project identifying all the parties involved who will need formal project communications and pressing these parties to define their communications requirements. Of course there will be some modifications to the plan as the project progresses, but the bulk of the parties needing formal communication should be identified up front. Psychological and political barriers can be addressed with the outside world in much the same manner as within the team; the difference being that the project manager will have very little authority to make changes to the psychological or political landscape in the client world. The project manager should try to ascertain the psychological profile of the project's main clients and then use that information to determine how best to communicate with those clients. It also behooves the project manager to ascertain the political climate of the client community and try to determine early on where alliances exist between clients. Quite often client behavior seems to have no rational motivation until you understand the underlying political landscape, and then it all makes perfect sense. I have seen a number of project leads needlessly pound their heads against the wall trying to make sense of their clients, only to find out later that political machinations were at work behind the scenes. Had they been able to dig these political motivations out up front, they could have made their jobs much easier and could have done a much better job of satisfying the true desires of the clients (Table 5.3).

This is a small sample of a project communications plan that would be incorporated into the larger total project plan. The main items to include

TABLE 5.3   Communications Plan

Person	Type of Communication	Medium	Delivery schedule
Client manager	Weekly project status	Conference call	Every Thursday at 1 P.M.
Project stakeholder	Monthly project status	Written report	5th of every month by 10 A.M.
Project manager	Daily update	E-mail	Daily by 4 P.M.
Project team member functional manager	Weekly project status	E-mail	Every Thursday by 3 P.M.
Project team member	Daily status update	Face-to-face meeting	Every work day at 9 A.M.

in the plan are the personnel who will be expecting communication from the project, the type of communication these people have specified that they need, the medium in which the information will be delivered, and the schedule for information delivery. Other information may be added to the plan as needed by the particular project, but this should be the minimum basic information included on any project communications plan.

Conclusion

Efficient communication is essential to the success of any project team, but even more so in the high-tech world where the business climate changes so rapidly and decisions have to be made quickly. There have been many volumes written on the subject of effective communication. You can get into much detail dealing with sensors and receptors, implied meanings, body language, and the like. Effective communication, though, really comes down to a few simple, commonsense guidelines:

- *Say what you mean.* Be precise and concise with your language. Do not say any more or any less than is necessary to convey your thoughts or to relay information on to another person.

- *Know the person you are trying to communicate with.* Try to obtain as much information as possible about the personality of the person with whom you are trying to deal. Do they like their information delivered face-to-face or do they prefer it to be delivered via e-mail? Know what information the other person needs to have communicated to them. Don't deliver an in-depth tome of material when a brief overview will do. Finally, determine the political power of the person you are dealing with and the political ties they have with both your project team and your client community.

- *Use the right communication vehicle for the situation.* Use informal methods of communication for short, quick bursts of communication. Use more formal methods for longer communications, scheduled periodic communications, and communications over long distances or to large groups of people all at one time.

Effective communications boils down to saying (or relaying) the right information at the right time to the right people in the right manner.

6

Motivation systems for technical teams

Introduction

We have seen the importance of communication systems for project teams. Now let's take a look at communication between the project lead, or even the whole organization, and the team members. In particular, we'll look at how team members can be influenced to perform the project tasks assigned to them. One of the most important jobs of the project leader is to motivate his or her project team to perform at their peak level of performance. The word motivate comes from the Latin word *motus*, which means *to move*. A project manager must move his or her team to put the necessary energy into their project duties so that the goals of the project may be accomplished within the triple constraints.

Motivational theory

Many theories on motivation exist. Let's briefly discuss a few of these theories so that we might get a more complete picture of what motivation is all about:

- *Goal-setting theory*. This theory works off the premise that the characteristics of the goals a person pursues impacts the amount of energy that person will expend in pursuing those goals. The theory was developed by Edwin Locke in the 1960s as a result of work Locke was doing to discover what influenced people's behavior on the job. He believed, and later research supports, that people will work harder when they have goals than when they don't. The goals should be challenging and specific, they should be time-bound and reasonable, and they should be measurable and rewarding. Furthermore, the individual must accept the goals put forth and must have the opportunity to give and receive feedback on the progress they are making towards the goals. Finally, people tend to work harder towards goals they had some influence in developing. It pays to include the project team when goals for the team are set.

- *Reinforcement theory*. Reinforcement theory is supported by behavioral psychologists who believe that a person's behavior is directly tied to the consequences that behavior produces. Behaviors that produce rewarding consequences will be repeated and behaviors that produce unrewarding or unpleasant consequences will become extinct. Punishment for a behavior may cause that behavior to be suppressed, but the behavior may return when the memory of the punishment fades or when the threat of the punishment disappears. Rewarding consequences can be divided into two groups—intrinsic and extrinsic. Intrinsic rewards are generated internally and are a natural result of performing tasks in a certain manner. Satisfaction with a job well done is an example of an intrinsic reward. Extrinsic rewards come from external sources and are typically more tangible types of rewards such as increased pay or formal recognition for performance. Extrinsic rewards may also include such intangibles as a change in a working relationship with a superior or inclusion in high-level decision making. Extrinsic rewards should be used to motivate behavior that would not be motivated intrinsically under normal working conditions. Project leaders must take care to dole out rewards judiciously. Team members appreciate simple thank yous and "attaboys" or "attagirls"; however, the project leader should not go so far overboard with rewards that they become expected. Sometimes extrinsic rewards will replace naturally occurring intrinsic rewards and then the project leader is stuck continuing the flow of extrinsic rewards to keep the team's motivation at an acceptable level.

◆ *Equity theory.* Equity theory deals with people's perceptions of both the absolute and relative values of the rewards they receive. People will compare their situation with previous situations they are familiar with and also with the situations they perceive others to be in. No matter how great the absolute value of the reward seems to be, if a person does not perceive the relative reward to be what he or she feels is equitable, that person will not feel motivated to perform at their peak level of performance. The reverse is sometimes true. If a person perceives themselves as being overcompensated for their work, they tend to work harder to make up for it. Typically though, equity theory deals more with the perception that a person is undercompensated rather than overcompensated.

◆ *Expectancy theory.* Expectancy theory, developed in the 1960s, deals with three elements of a person's perception of a task: (1) the perceived ability to accomplish the task, (2) the perceived chance of receiving a reward for performing the task, and (3) the perceived attractiveness of the reward. Motivation to perform a particular task is based on the product of these three factors, which means that if any one of the factors rates a zero, total motivation to perform the task will be zero. Expectancy theory brings to light the fact that a project manager must match project tasks to the project team members who possess the skills to perform those tasks most efficiently. It also places a burden on the project manager to understand what rewards will motivate members of the team, realizing that what motivates one member may or may not motivate another. This last item brings to light a situation I have seen repeated many times in several different organizations. A committee is formed to design an awards program to recognize employees for good work or extra effort or whatever the case may be. These programs typically do not take into account the differences in personalities between work groups, functional groups, or individuals. A common set of awards is developed regardless of their motivational appeal to individuals. I have actually seen these award programs become demotivating to employees, particularly in the technical arena. Technical project members tend to thrive on challenge and the opportunity to chart their own career choices. Tokens of appreciation such as plaques and certificates are not great motivators for technical people. I know from my own personal experience that this is true and I have confirmed my suspicions with informal polls of my colleagues over the years. Rewards must be tailored to individuals in order to achieve maximum motivational benefit.

Now that we have explored several theories on motivation, let's take a look at some specifics forms of motivation, particularly as they relate to the technical project team.

Monetary compensation

Monetary compensation in technical fields such as engineering and software development has been a hot topic for some time now. Demand has outstripped supply of qualified personnel in the technical arena for years and, as a result, many companies have had to increase the amount of monetary compensation offered to people skilled in technical disciplines. Monetary compensation may come in different forms. It may be in straight salary, it may be in incentive bonuses, it may be in stock options, or it could be in any number of other forms. Most organizations expend considerable resources benchmarking their pay practices against competitors in the industry and against other industries. Monetary compensation is definitely an important bargaining tool for organizations trying to recruit top technical talent. The problem many organizations run into arises out of the fact that top management often sees monetary compensation as the most important factor affecting job satisfaction while front-line workers place it farther down the list. Certainly monetary compensation is important. If you look at Maslow's hierarchy of needs , you will find that physiological needs are the first needs humans will try to fulfill (Figure 6.1). Monetary compensation is the key to fulfilling these needs. As you move up the pyramid of needs, however, you find that monetary compensation has little or nothing to do with meeting those needs. Therefore, monetary compensation can only be a starting point for attracting and retaining technical talent. It would be impractical to imagine that any one company could outbid all its competitors as far as salary goes, and even if an organization could do this, it would not have the resources left over to offer other incentives to attract and retain employees.

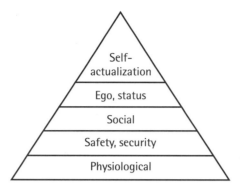

Figure 6.1 Maslow's hierarchy of needs.

Supply and demand in the marketplace also plays a part in the monetary compensation policies of organizations. Almost everyone knows someone who benefited from the demand for technical personnel in the years leading up to the year 2000. Demand for technical services to deal with the Y2K problem far outstripped supply and people with even marginal skills were able to reap huge monetary rewards. Now, only months after the turn of the millennium, the demand has all but dried up and positions that once commanded astronomical sums of money not only don't pay that well anymore, they are hard to find even at reduced wages.

Awards and recognition

If you examine the literature on employee motivation, there are volumes written about the importance of awards and recognition. Some authors would have you believe that the more awards you present to your team and the more recognition you heap on them, the more motivated they will be to do a good job. This may be true to a point, but my experience in the technical world leads me to another conclusion. I do not deny that we all need to know our work is appreciated. As I have stated earlier in this chapter, it is easy to get so carried away with awards and recognition that they become meaningless. Couple this with the fact that technical personnel generally place challenging work, responsibility, and monetary compensation ahead of awards and recognition and you can see that the project manager of a technical team must have an understanding of these issues or he or she will do more harm than good with a recognition program. It can be more difficult than you might first imagine to actually align a reward program with the goals of the organization. Allow me to illustrate this point with two examples from American military experiences. Prior to the conflict in Vietnam, soldiers were sent to war for the duration. Thus, once the war was won the reward of coming home was gained. In Vietnam, however, soldiers were sent into the conflict for a specified period of time, so the motivation of winning the war was removed and replaced with personal survival for the specified period of time. The lessons learned from this conflict were applied successfully in the Persian Gulf. Personnel were sent to the Middle East to accomplish a goal, not just to survive. The results are markedly different when you match your reward system with the goals of your project. Another aspect to consider when contemplating awards and recognition for a project team is the power to bestow these rewards on project members. The project manager may not have the authority to reward his or her team. That authority may lie with the functional managers in the organization. In situations where the project manager does not have the authority to formally reward his or her team, creative solutions must be developed.

Creative motivational methods

What's a technical project manager to do, especially when he or she does not control the purse strings? First, the project manager should pay heed to what history has said about reward systems. Lao Tzu, the sixth-century-B.C. Chinese philosopher, said, "Reward for merit brings strife and contention." More recent thoughts come from W. Edwards Deming, who said, "The present style of reward squeezes out from an individual, over his lifetime, his innate intrinsic motivation, self-esteem, dignity. They build into him fear, self-defense, extrinsic motivation. We have been destroying our people from toddlers on through the university, and on the job." Finally, lecturer Alfie Kohn had this to say: "The introduction of, say, monetary reward will edge out intrinsic satisfaction; once this reward is withdrawn, the activity may well cease even though no reward at all was necessary for its performance earlier. Extrinsic motivators, in other words, are not only ineffective but corrosive. They eat away at the kind of motivation that does produce results." The basic thrust of all these ideas is that organizations should be wary of introducing reward systems that will replace intrinsic rewards for extrinsic ones. The project manager who does not control the purse strings is in a unique position when it comes to reward systems because he or she cannot introduce the damaging extrinsic rewards mentioned earlier. Of course, the project manager is also faced with the task of creating inventive rewards that will motivate the team without appearing shallow and insignificant. One way to reward technical team members is to continue to challenge them once they have proven themselves up to the task. Give them increasing responsibility over their work and the decisions that need to be made in the course of the project. By showing your team members that you trust them and need their contributions you will help generate the intrinsic rewards of the job that are most satisfying. Other creative ways to reward the technical project team might be to select a prime work location for your team. Negotiate for convenient work locations, locations with lots of windows, locations that are close to shopping and dining areas—whatever your team indicates that would motivate them. If you have remote work locations that would also be fun travel destinations, use this as a motivating technique. Reward good work by placing employees in these locations. Provide training opportunities for your technical team members, and again, if the training can be scheduled in a fun location, schedule it there. None of these solutions involve spending tremendous amounts of money, but I'll guarantee they will go farther in satisfying technical project members than any award or even token monetary amount you could offer. You might even go so far as to gather the project team for a brainstorming session to determine the types of rewards that would do the most to motivate each of the team members. The results might be surprising. One example of creative motivational methods comes from Eastman Kodak. This company employs a performance commitment plan which ties performance pinpoints and the behaviors used to achieve those pinpoints to a system of point scoring.

> For mutually agreed WBS tasks and/or summary tasks, the team established with the program manager a number of points to be awarded for production of the deliverable. No deliverable, no points. As the projects unfolded, there were occasional situations which made it clear that subsequent deliverables were no longer viable, and these were renegotiated and the associated points were reallocated to new, remaining deliverables. This situation accounted for less than 10% of the total points. The project management consultant who supported this program has over twenty-

odd years of project experience. He reported he had never seen a program in which people did as good a job of planning the work, then working the plan. Project team members generally responded enthusiastically to the challenge to be measured by the results of their project deliverables. Deliverables were challenging. Performance commitment plans linked to pay worked for a great majority of contributors. [1]

Quality of life

I have seen a dramatic shift in the attitudes of employees in technical disciplines towards their employers over the last 20 years. The days of the company man are fast fading away, if they haven't gone already. Employees have found that when organizations are faced with what is good for the organization and what is good for any one particular employee, the organization wins every time. I have seen instances of employees uprooting their families and moving from their home territory halfway across the country just to keep their job with the organization, only to have the organization lay them off six months later. Employees are coming to the realization that they must take care of themselves, because the organization is not going to do it for them. Consequently, employees are more likely to look for another job rather than move with the organization if moving means sacrificing the quality of life the employee and his or her family currently enjoys. Again, intrinsic values are winning out over extrinsic ones. Some organizations are beginning to realize that in addition to monetary incentives and recognition, they must also offer quality-of-life incentives if they want to attract and retain the top technical talent available. These incentives might include, but would not be limited to, onsite child care, onsite fitness facilities, company-run health care facilities, a generous vacation policy, flexible work schedules, and the opportunity to telecommute. Employees are demanding that they have more time with their friends and families. They are demanding more choice over where they are able to live and work. They are demanding support for preventative health care measures. In short, they are demanding more control and more balance between their working lives and their nonworking lives. Project leaders must be in tune to quality-of-life desires from project team members and use this knowledge when project schedules are built and resources are allocated to a project. Organizations have the ability to offer quality of life incentives on a large scale.

Project leaders can also address the quality of life issue by allowing for flexible work hours and flexible work schedules, managing projects in such a manner that team members aren't caught working 80 hours per week to bring a project in on time, and by distributing the work load equitably among team members. My experience in the technical arena indicates that quality of life issues will drive employees away from an organization more quickly than any other motivational issue. If employees aren't happy with their life outside of work, it will impact their life inside of work. Their motivation to perform at their peak level will be low and the quality of the work they produce will not be acceptable.

Inspiring innovation

Motivated employees are innovative employees. The tenets of total quality management call for continuous quality improvement, but it is in the breakthrough moments that organizations are able to leap ahead of their competition. Innovation is necessary for organizations to remain viable as the business climate changes. In the high-tech world the business climate can change on a dime, making innovation among employees even more critical. Project leaders must be aware of every source of motivation available to them. They also must understand what their team members need in the form of motivation. Project leaders must realize that motivation for technical teams is different than for other types of teams. Technical personnel typically view the world from a more pragmatic standpoint, thus motivational systems for technical personnel must be rooted in this same world view. Token rewards such as plaques, certificates, and the like can be demotivators in the technical arena. Technical personnel often take the viewpoint that management is not willing to provide true motivational rewards. They don't have the confidence in their personnel to reward them with more challenging work or increased decision-making authority, so they substitute meaningless tokens of appreciation. Project leaders should guard against this situation and should work to provide the proper motivation for their technical team members.

Reference

[1]   Hellawell Jr., G., *Team Development Meets the Performance Based Culture,*
Newtown Square, PA: Project Management Institute, 1997.

7

High-tech project teams and quality

Introduction

Teams that communicate well and are highly motivated can be expected to turn out quality work. This chapter will take a brief look at what we mean by quality and some techniques to help teams achieve a desired level of quality in the work products they produce.

A brief history of the quality movement

The demand for consumer goods produced in the United States skyrocketed after the end of World War II. Consumer goods had been rationed in the United States during the war, and neither Japan nor Europe was in a position to resume production of such goods given the devastation they had both experienced during the war. Any products U.S. factories could produce were quickly snapped up by the worldwide consumer community. This was a boon to the bottom line of U.S. industry, but it also led to complacency and inefficiency by U.S. corporations. They had no incentive to do things faster, or smarter, or better because they had no serious competition anywhere in the world. Productivity and quality in the U.S. manufacturing sector dived as quickly as profits soared. At the same time, Europe and Japan were beginning to rebuild their industrial bases. Given their situations, they saw a need to embrace quality. They saw the benefit of working smarter, not harder. They were starting their industrial production from scratch and could embrace the concepts of quality easily. Through the work of W. Edwards Deming and Joseph Juran, the Japanese in particular rebuilt their manufacturing sector based on the tenets of total quality management (TQM). Although their success in quality management did not happen overnight, by the early 1980s, the United States realized the predicament it was in with regard to industrial production and manufacturing. Japanese products were in great demand in the United States because they were simply built better than their U.S. counterparts. It became apparent to even the highest ranks of U.S. industry and government that something had to be done and soon (Table 7.1). That something was kicked off with the development of the Malcolm Baldridge Award for Quality. The purposes of the award are as follows:

- Stimulate U.S. companies to improve quality and increase productivity;

- Recognize those companies that improve their quality and increase their productivity and hold those companies up as examples for the rest of U.S. industry;

- Establish criteria that U.S. business can use to evaluate internal quality-improvement efforts;

TABLE 7.1   Quality-Control Methods in the United States

Pre-1900	Operator quality control
1910s	Supervisor quality control
1920s, 1930s	Quality control by inspection
1930s–1950s	Statistical process control
1950s–1980s	Quality assurance systems
1980	ISO 9000 Technical Committee TC-176
1983	TQM principles established
1987	Malcolm Baldridge National Quality Award established
Future	Kaizen

- Publish the strategies of winning companies so that the rest of U.S. industry might use these strategies as a guideline for improving quality and increasing productivity.

The main thrust of the current TQM effort in the United States is the elimination of waste in processes. This can include wasted scrap material, wasted time in rework, and wasted dollars in financing redundant or unnecessary processes. The fundamental tenets of the current TQM movement can be summarized as follows:

- *Quality is defined by the customer.* The customer can be an external end user or just the next person in the internal production line.

- *Teamwork.* Production processes must be viewed as interdependent from the input of raw materials into a process to the final output of the finished product.

- *Mathematical metrics.* Product quality must be measured during the production process using statistically valid means. Quality must be built in. It cannot be inspected in.

- *Kaizen.* The Japanese term for continuous quality improvement is kaizen. Quality improvement does not occur in large leaps, but small incremental steps.

- *Benchmarking.* Why try to reinvent the wheel with your production processes? Find the best of the best in your industry or in your process group and emulate them. Then build on this base of quality to achieve kaizen.

Planning for project quality

Quality does not just naturally happen in the workplace. An organization must develop a plan for continuous quality improvement in all the products or services it produces, all the processes utilized within the organization, and in the work ethic of the employees. Planning also includes communicating the requirements for quality in a manner that is understandable and complete. A first step in the development of this plan is to perform a quality audit of the organization. Deming provided a checklist of points to review and to compare against when an organization performs such an audit the following:

1. *Policy.* Dissemination of and execution of the policies for management, quality, and control should be examined. The appropriateness and clarity of presentation of these policies should also be examined. Policies should be consistent, should tie to long- and short-term objectives of the organization, and their results should be statistically evaluated.

2. *Organizational structure.* Definition of scope of responsibility and authority should be clearly defined for all positions. Cooperation among all elements of the organization should be examined. Utilization of quality control teams and methods should be evaluated.

3. *Educational dissemination.* Both formal training and on-the-job training procedures should be reviewed. The application of this education should be measured and evaluated.

4. *Quality information.* The speed of collection and dissemination of quality information from inside and outside the company should be measured. The speed at which this information makes its way to the lowest levels of the organization, and the speed with which this information is incorporated into the daily workflow of the employee body should also be examined.

5. *Analysis.* The efficiency with which key contributors to waste are identified, analyzed, and dealt with must be measured. Critical problems must be analyzed within the framework of the existing technology and statistical measures should be employed to analyze both the problems and the solutions implemented.

6. *Standardization.* The method of establishing, revising, and abolishing standards should be examined. The contents of the standards should be reviewed and the utilization of the standards by the employee body should be measured.

7. *Quality systems.* Systems for the control of cost and quality should be examined. The use of quality-control measures, particularly statistical methods, by the employee body should be measured. Participation in and results of quality teams should be measured as well.

8. *Quality assurance.* Each step of an organization's production process, regardless of whether the organization produces goods or
   services, should be reviewed for inclusion of quality-assurance
   procedures. The utilization of statistical quality-assurance procedures along with the appropriateness of the tools utilized to assure
   the quality of organizational processes should also be examined.

9. *Future planning.* A measure of the grasp of the present state of af-
   fairs as they relate to the quality of processes within the organization should be developed. The plan for further improvement in
   quality processes should be reviewed as well as the plan's link to
   the long-term goals of the organization.

Once a quality baseline has been developed, the organization must on a
periodic basis measure itself against its quality checklist to determine
whether it is improving its level of quality or slipping below the baseline.
Organizations must always remember that quality has to be gauged based
on customer perception. No matter how well the organization thinks it is
doing based on its internal yardstick, customer perception is the true
benchmark. For example, quality in the telecommunications industry
has been measured for years against numbers like troubles-per-thousand
access lines, or seconds spent by an operator per call, or even the number of
employees per access line. While these are all good numbers to have on
hand, and they all certainly may be measures of some aspects of the quality
of service provided by the organization, they mean absolutely nothing to
the customer who finds himself or herself without dial tone. Customers
tend to judge telecommunications companies on measures such as how
polite the service rep was when they called for new service; how quickly
someone repaired their service once they reported trouble; and even how
often they find erroneous charges on their bills? One major telecommunications company, on its journey through the establishment of modern
quality control, decided that in addition to measuring success based on its
internal quality metrics, it would establish metrics based on customer perception. This company polled its customer base to determine what aspects
of service were most important. It then developed a survey with accompa-
nying metrics based on the responses to this poll. A statistically significant
random sample of the company's customer base was chosen and the survey
was administered to the members of this sample group in order to establish

baseline responses to the survey questions. Customers were asked to evaluate how their ideal telecommunications company would rate in the survey as well as how this telecommunications provider would rate. Going forward, on a quarterly basis, another sample of customers was surveyed to gather current responses to the survey questions. Each time a survey was conducted, the new numbers were compared to the baseline and the gap between this company and the perception the customers had of their ideal company was measured. The objective of the company was to close the gap between themselves and the ideal as perceived by the customer base. The survey allowed this company to better determine what was really important to their customers, how close or far away they were in meeting customer expectations, and what issues the company needed to address with process improvement efforts.

Knowing the quality issues to address in order to improve customer satisfaction is only one piece of the quality-planning puzzle; organizations must align their personnel both physically and mentally to address these issues in the framework of a total quality management program. I participated in the TQM program of a company for over 5 years and had the opportunity to watch the program evolve from a process managed separately from the employees' daily routine to one that was ingrained into each employee's daily routine. The program was originally based on Deming's quality principles and the CEO of this organization actually visited Japan and met with industry leaders there to learn about the TQM programs they had implemented in their companies. The first step the organization took in implementing TQM was to form quality circles. Quality circles were small groups of people who banded together to brainstorm lists of quality issues that needed attention, and then to develop solutions to these issues. Participants in a given quality circle were generally from the same work area, but some cross-functional quality circles existed. The groups typically had six to eight members plus a facilitator who had been trained in meeting facilitation and quality management. Each team member was also required to attend training on quality management, which included brainstorming techniques, root-cause analysis, team building, and quality improvement implementation. Quality circle members participated in team building activities such as team name development and team logo development as a way to develop camaraderie and to instill pride in the team. The logos were even framed and hung in the boardroom of the organization as a way to communicate to the employees the importance executive management placed on total quality management. The quality circles were a huge success. Quality circles were responsible for a multitude of process improvements that ranged from improved methods of cable splicing in the field to reducing the number of print lines in the IS shop. Both time and money were saved within the organization, but more importantly, the idea of quality was brought into the forefront of each employee's mind. Over time quality circles evolved into quality teams which were groups of employees from the same workgroup. The organization was pushing the evolution of quality management from a process which was adjunct to the daily workflow into one which was integrated into the daily workflow. Today this organization has taken a further step in the evolution of their TQM program by eliminating formal quality teams. Organizational and natural work teams are expected to employ proven quality principles into every aspect of their daily work. When quality issues arise, these teams perform root-cause analysis, they develop irreversible corrective actions for problems, they perform risk management, and they follow the steps outlined by TQM to implement solutions. The future of TQM in this organization is a continued march toward kaizen, whereby each employee looks for opportunities to make incremental quality improvements in their everyday job processes. Stay tuned!

The cost of quality

The initial reaction of management about to embark on quality implementation is, How much is this going to cost me? They often see quality implementation as a major change in the way they currently run their business and subsequently believe that it will be an expensive proposition. This may or may not be true. If a company currently performs well, competes effectively, and satisfies its customers, the implementation of a quality program may be as simple as fully documenting existing company policies and putting metrics in place to insure that the company operates within those policies. According to Paul Gladieux, "Companies that perform well typically have 90% of the core information required for implementing an effective quality system" [1]. Of course, if you are starting from ground zero the cost to implement a quality management system will be much higher, but few companies, other than ones just getting started, will be in this situation. Every organization has to have some sort of quality management and quality control in order to compete, regardless of the market. There is a cost to quality regardless of whether it is for a TQM program or whether it is for the rework that occurs because no quality management program is in place. According to Lewis Ireland in *Quality Management for Projects and Programs*, "Currently, many projects are not measuring the cost of nonconformance activities. The few that have measured the costs of quality reveal that several projects have extra costs in excess of 20% of the total cost of the project because of the lack of quality procedures. In many recent projects, 12% to 20% of the project costs can be attributed to waste. This does not include those projects which were never completed or those that had cost overruns in excess of 100%. The typical project should have a goal of between 3% to 5% of the total value as the cost of a quality program, depending upon the type of project and its total dollar value" [2].

Management's next question may be, "When will I see a return on my investment in quality management? Again, the answer will vary depending on what point you start from, but if we accept Gladieux's estimate, the return on investment (ROI) for quality implementation should be sooner rather than later. Costs of quality implementation can be viewed in two broad categories: startup and maintenance. Startup costs could include project planning, human resource management, and training. Maintenance costs include such things as procedural revisions, data gathering, inspection, and audits (Table 7.2). Most of these costs are quantifiable, but the real problem lies in how to quantify the results of your quality implementation. How much money do you save by only executing a process once as opposed to twice? How much money do you save by identifying a defect in a product at step two of a process as opposed to step three? These costs are not quite as easily stated in monetary terms. One way to try to capture the cost savings of quality implementation is to compare the time spent in each step of a process as it exists today against the time spent in each step of a process after you implement your quality program. The method I have used to document the before and after processes is simply to flowchart the process that I am dealing with at the time. This requires interviewing each participant in the process to document the steps in the process and to assign a time value to each step. You must also be sure to document any wait time within the process, such as when a form sits on someone's desk awaiting a signature. I have participated on a number of quality-improvement teams that have performed this exercise, and I can attest to the effectiveness of this approach. The results are often shocking

TABLE 7.2   Quality Implementation Costs

	TYPE	EXAMPLE
Startup costs	Project planning	Development of detailed plans to establish and implement quality teams within an organization
	Human-resource management	Development of HR policies detailing quality team activity
	Training	Quality team facilitator training
		Quality team leader training
		Quality team member training
Maintenance costs	Procedural revisions	Updating and distributing policy modifications
	Data gathering	Periodic reporting of process metrics
	Inspections	Periodic internal inspection of processes for compliance with stated procedures
	Audits	Formal third-party audits to check for compliance with stated procedures

because they present the cold, hard truth about the inefficiency of the current process you are looking to improve. As an example of this, one team I worked with was presented with the problem of improving the process by which our clients requested work to be performed by the IT development staff. The old process required that our front-line clients complete a paper form which then had to be signed by two supervisors in the requesting department plus have an account code assigned by a particular group in the accounting department. Once all these signatures and approvals were obtained, the request was forwarded to a clerk in the IT department who manually logged the request into an on-line system. The clerk then forwarded the request to the IT development manager for another approval and for assignment to an IT development supervisor. The supervisor then assigned the request to a developer to be worked. The whole process could sometimes take two weeks, and this was just to get a request from a client to a developer who could actually do the work! The client departments and the IT staff knew this request process was inefficient, but it really hit home with all of us when we saw the process laid out in black and white. An interdepartmental team was formed to revise the process. Each step was examined to determine if it was even necessary. Wait times were inspected and eliminated when possible. The improved process eliminated one client signature, the account code assignment, and one signature in the IT department. As a result, the two-week process became, at most, a two-day process. The impact of this improvement becomes glaringly apparent when you subtract two weeks' time from the hundreds of requests that came into that IT development group every year, not to mention the fact that the time of two managers and a clerk in accounting was saved as well. The time and dollars invested in this project were returned to the organization in the space of less than 6 months with the ongoing savings there to be reaped for years to come.

So far we have discussed the price of conformance to a quality program. What about the cost of nonconformance? What is the cost to the organization if it chooses not to implement a quality management program? What is the cost to the organization if its employees do not work within the guidelines of the quality management program? The cost of nonconformance is directly proportional to the loss of profit on your product or service that you would realize if you had to price that product or service at the same level as the industry leader given equal functionality of the products or services being compared. You can be assured that the industry leader, regardless or product or service, is practicing sound quality management. You cannot expect to be competitive in today's business climate if you do not employ quality management principles. The cost of nonconformance will either force you to price your product or service above what the market will bear, or the quality of your end product will be so poor that your sales will not support your production expenses. The cost of nonconformance may include the following areas:

- *Waste of time and material.* Time and material is wasted when poor-quality work has to be redone just to meet minimum standards. Manpower and productivity are lost. Materials may have to be scrapped, or extra material may have to remain in inventory to cover rework, which drives up material and material-storage costs.

- *Schedule slippage.* The more inefficient a process is, the more chance there is that its delivery schedule will be unpredictable. Payment for work that is not completed may be deferred and the cost of the money that would have been paid can erode potential profits on the project.

◆ *Image.* The public image of a product or service may determine subsequent purchases of that product or service. A poor image may
result in reduced sales volume and increased costs dealing with complaints. Quality work cannot guarantee future sales, but lack of quality can guarantee an adverse impact on sales.

In short, the eventual cost of nonconformance is extinction.

Tools to measure quality

Modern quality managers employ a number of tools for both quality assurance and quality control. Some of the tools require knowledge of statistics
while others are simple visual methods.

◆ *Statistical process control.* Statistical process control can employ both
process control charts and acceptance sampling. Process control
charts are a statistical technique to monitor a process in progress
while acceptance sampling is a technique used to measure the results
of a process that is already complete. Before either of these methods
may be used, standards of acceptance must be established. These
standards may be based on internal measures of acceptability or they
may be based on measures dictated by the customer. Process control
charts make use of a normal statistical distribution to determine an
upper and lower control limit. The average of each iteration of the
process, or small samples of the process, will determine the mean of
the control chart. Once the chart is established, results from each
run of the process are plotted on the chart to produce a visual representation of the status of the processes over time. If a process result
falls outside the range of the upper and lower control limits, that run
should be investigated immediately. In addition, if seven or more
points on the chart trend in the same direction, the process should
be investigated because this indicates that the process may be headed
out of control. Acceptance sampling, on the other hand, deals with
each individual run of a process as it happens. Testing every single
output from a process run is expensive, time consuming, and, in
most cases, not practical. Statistical methods are used to determine a
sample size that will replicate the results of the whole lot. Inspection
criteria have to be developed and a maximum number of defects

acceptable must be established. In this manner the output from a process run will either be accepted or rejected based on the results of the acceptance sample inspection.

- *Pareto diagrams.* Pareto diagrams make use of histograms to rank the number of defects by order of frequency of occurrence showing 100% of the defects. They allow you to see which defects occur most often, and subsequently, which defects need the most immediate attention. By attacking the defects which occur most often, you can bring the total number of defects down more quickly than if you just attack the defects randomly. Other considerations should also come into play in ranking defects on a Pareto diagram. It may be that the most frequently occurring defect is not the one that is causing your customers the most pain or the one that is costing you the most money.

- *Cause-and-effect diagrams.* These diagrams provide a structured means of performing root-cause analysis to determine the true nature of a problem or defect, so that we can fix the right problem. Typically there are eight major inputs to the diagram which include time, energy, machine, measurement, method, personnel, material, and environment. Questions to ask about the defect should always include who, what, where, when, why, and how, and then an answer from each of the major inputs should be explored for each question.

Other simple tools for quality assurance and quality control include scatter diagrams, histograms, graphs, checksheets, and checklists. These all provide visual representations of process outputs and time-scale trends of process outputs.

Conclusion

Quality management has been embraced worldwide. One organization cannot decide for itself anymore the level of quality at which it will operate. The world consumer market sets the quality standards for industries today. Organizations must take the following steps if they are to survive:

1. *Educate themselves on the main principles of quality assurance, quality control, and total quality management.* Senior managers should

study the works of Deming, Juran, and other icons in the field of TQM. They should familiarize themselves with the requirements for the Baldridge National Quality award, the Deming prize, and the standards set out in the ISO 9000 documentation. Once upper management has been educated, quality training should be provided at all other levels of the organization so that every employee in the organization can speak about quality intelligently.

2. *Plan for quality.* Just as every project needs a plan to provide direction and to enumerate the steps that need to be taken in order to achieve a goal, an organization must have a plan to achieve a higher level of quality in its outputs. This plan must include the quality standards for the organization as determined by the customer base, the metrics to be used to measure quality within the organization, and a method to maintain the level of quality within the organization as the business climate changes and customer needs change.

3. *Achieve kaizen.* Companies should never give up in their quest for quality. They should continue to preach quality to the employee body, they should continue to adjust their quality plan to meet the needs of the customer base, and they should encourage and empower employees to make small, daily, incremental improvements in quality. Once an organization has reached a level where every member of the organization looks to make quality improvements every day, they will be in a state of kaizen. The goal of every single employee should be to do the right things right the first time.

References

[1]   Gladieux, P., "Quality Management System Standards," *PM Network,* February 1995.

[2]   Ireland, L. R., *Quality Management for Projects and Programs,* Newtown Square, PA: Project Management Institute, 1992.

8

Negotiation strategies for high-tech project managers

Introduction

We'll shift gears somewhat now and look at some topics that deal with the project teams' interaction with the outside world. Our first topic will be negotiation strategies. We'll start with some basics then move on to techniques to use depending on the negotiating situation you happen to find yourself in. Finally we'll talk about possibly the most important negotiation skill—listening.

Negotiation basics

Communication is undoubtedly the key skill a project manager must possess if he or she is going to be successful. One use of those communications skills will be negotiating with stakeholders, sponsors, and team members. We all negotiate every day and begin learning our basic negotiation skills early in life. Negotiation is a basic means of getting what we want. Negotiation is back-and-forth communication designed to reach agreement when you and another party or parties have both shared and opposing interests. Many styles of negotiation exist and it is important to know who you are dealing with and what the motivations are behind each party to the negotiation before you pick a style. There are some negotiation basics, however, that will assist you in negotiating more efficiently and more effectively:

- *Avoid establishing irreversible positions.* Typically when we negotiate, we choose a stance on the issue at hand, then haggle back and forth until we meet somewhere in the middle. Think about the last time you purchased an automobile. You, as the buyer, have a particular make, model, feature set, and price that you desire. The seller has a price that he or she is asking, and also a minimum price that he or she will accept. The asking price represents the seller's opening stance while the minimum acceptable price represents how far off that opening stance the seller is willing to move. If the buyer is unwilling to come up to that price and if the seller is unwilling to drop below the minimum acceptable price, no deal will be made and neither party will walk away happy. In the same manner, project managers should avoid digging into positions before negotiations have begun. Timelines, budgets, and scopes cannot be reasonably set until all parties with a stake in the project have had the opportunity to express their needs. The project manager must keep an open mind at this point in the process and guard against digging into a position that he or she will have to retreat from later. This type of behavior will create credibility problems, from which the project manager will have trouble recovering, if he or she can recover at all. By the same token, a project manager should avoid agreeing to a position taken by another party in the negotiation just for the sake of reaching agreement. Clients are notorious for setting end dates for projects before the scope and the project tasks are defined. Sometimes these end dates are true drop-dead dates, but most often they are not. A project manager only creates problems for himself or herself if he or she agrees to arbitrary project end dates. Once he or she agrees to such a date, it is very difficult to convince the client that a legitimate reason exists to move the date out. Again, the project manager only creates a credibility problem for himself or herself by working from or with negotiating positions.

◆ *Separate personalities from negotiations.* Parties to a negotiation are still human beings with personalities, preferences, and feelings. If you've ever been involved in a negotiation where personalities became involved, you know how messy the situation can get. It is impossible to negotiate personality differences because there is no right or wrong when it comes to thoughts, preferences, and emotions. There is no room for negotiation on these issues; therefore, as a project manager, you must stay away from these areas. The difficulty comes in when you consider that any negotiation generally involves a substance level and a personal level. Each party to the negotiation obviously wants to achieve his or her stated objectives, but each party is also trying to establish some sort of personal relationship with the other parties. This relationship may be for the benefit of the current negotiation or it may be in anticipation of future negotiations. Whatever the rationale, personalities are tightly intertwined with the business portion of a negotiation and it is difficult, but necessary, to separate the two. Some ideas to consider when trying to separate personalities from negotiations are (a) avoid jumping to conclusions, (b) openly discuss each party's perception of the negotiation environment, (c) allow time away from the negotiations to release tension, and (d) refuse to respond to emotional outbursts. Failure to follow these suggestions will often lead to escalated emotional tensions if the negotiations don't proceed to everyone's liking, and once that escalation starts it is nearly impossible to stop.

◆ *Be creative.* Negotiations should not devolve into winner-take-all contests, nor should they produce mediocre compromises that leave everyone unsatisfied. A successful negotiator should look for solutions that make a winner out of all participants. In order to accomplish this, the successful negotiator must avoid a number of pitfalls. First, passing judgment on possible scenarios. Just as you withhold judgment when brainstorming ideas for potential projects, so should you withhold judgment when brainstorming ideas for solutions to a negotiation. Don't stifle creative problem solving by shooting ideas down before they have the chance to fly. Second, don't try to home in on the one right solution to the negotiation right off the bat. Too often we head down the first available path when searching for the solution to a problem and too quickly ignore other alternatives. Don't be afraid to let ideas flow freely and to allow for plenty of discussion. Third, think selflessly. What is good for the other parties in a negotiation is not necessarily bad for you. As you seek creative solutions to a negotiation, consider all parties involved, not just your own interests. Not only will this prompt your negotiation partners to act in a like manner, it may also open up solutions that you otherwise might not have discovered.

- *Rely on independently developed selection criteria.* Negotiations should never be based on subjective criteria, nor should one party to the negotiation determine the measuring stick to be used to determine what is fair and what is not. Universally recognized standards should be used whenever possible, such as IEEE standards, GAAP standards, and others. If these types of standards are not available, easily agreed-upon standards such as fair market value or legal precedent should be employed. The use of fair standards helps eliminate mistrust between the negotiation parties, which in turn leads to more open communication and cooperation between the parties. Objective standards should also apply and be applied equally to both sides. One party to the negotiation can't argue to use a set of criteria that they refuse to apply to everyone else.

Hardball tactics

So far we have looked at basic techniques to use when all parties to a negotiation are interested in dealing fairly and impartially with each other. Many times this is not the case. Let's next examine some of the tactics parties may use when they try to stall negotiations or try to win their way through means other than those discussed so far:

- *Use of unsubstantiated facts.* Parties to a negotiation may introduce statements to a negotiation as matters of fact, when they know that these statements are false. They may also make claims that they believe might be true or that they wish to be true but which have not officially been substantiated. Insist on verification of statements submitted as fact to a negotiation. This is not to say that you should distrust your negotiation partners. Rather, you should separate trust from the proceedings just as you separate personalities from the proceedings. Independent verification of factual information should be afforded every party to the negotiation, and in fact should be a part of the ground rules established when the negotiations begin.

- *Use of false authority.* All parties to a negotiation should be vested with the authority to act as agents for their organizations with full power to make binding decisions. Again, this verification should occur at the outset of the negotiations. If the parties to the negotiation do not have the authority to make binding commitments, they will have no incentive to negotiate openly and honestly. It is also a waste of everyone's time if a negotiator has to leave the table to ask permission to agree to a proposal.

- *Use of mind games.* Unscrupulous negotiators may employ a variety of mental tactics in order to sway a negotiation in their favor. These tactics may range from a mild omission of key facts to outright untruths. The unscrupulous negotiator may resort to threats, personal attacks, and even a refusal to negotiate at all. Mental tactics are extreme negotiating strategies designed to actually move away from negotiated settlements to settlements reached by force and deception. They are employed when one party is not willing to deal in good faith, or when that party is unwilling to seek out mutually beneficial settlements.

Preparation for negotiation

It is clear that successful negotiation requires a project manager to come to the table as prepared as he or she possibly can. What kinds of preparation will enable the project manager to achieve his or her goals from a negotiation? What steps should the project manager take ahead of time to successfully navigate through the negotiation process? Preparation is a key component of a successful negotiation experience. The following are some items to consider when preparing for a negotiation:

- *Past negotiation experiences of the other parties.* If the parties you are dealing with have had bad experiences in past negotiations, you will probably face resistance right off the bat. If you can collect data on previous experiences of your negotiation partners you will be able to gain insight into the positions they may possibly take in the current environment. In addition, you may be able to obtain information regarding the tactics the other parties tend to use in a negotiation setting. This will allow you to prepare your defenses in case the other parties try to use unethical or unorthodox tactics. You may also be able to discern which techniques work well with the others parties and which techniques do not.

- *Relative strength of the negotiating parties.* Sometimes the parties involved in a negotiation will be on equal footing, both from an authority standpoint and from the standpoint of the strength of the various negotiating positions of the parties. Most often, however, one party will have some type of advantage over the other parties. As a project manager, you need to be aware of your position in relation to the positions of the other parties. If you are in a position of strength, you need to know it. If you are bargaining from this position you may be able to stand firmer by your position than your negotiating partners. If you are in a position of weakness, you need to be prepared to yield more readily to your negotiating partners. However, you must determine how much you are willing to yield before you walk away from the table.

- *Focus.* You must know what it is that you want to get from the negotiation process before you ever start the process. Not only should you know your desired outcome, you should understand the true motivation behind your desired outcome. Developing a clear focus before the process starts will also serve you well if the negotiations drag on for an extended period of time. Sometimes you will be tempted to settle for less than you should just to bring the negotiation process to an end. If you have a clear focus of where you want the process to go and why, it will be easier to stand firm even when you would like to give in. Finally, a clear focus will allow you to articulate your needs clearly, which will lead to a more persuasive presentation to your negotiation partners.

Determining success

Once you have negotiated an agreement, whether it is a multimillion-dollar contract with a major supplier or a set of requirements for a project with a project sponsor, how do you know whether you have reached a successful outcome? Danny Ertel of Conflict Management suggests that there are seven points you should review to determine the success of the negotiated outcome [1]:

1. *Ensure that the negotiated agreement is better than the alternative, had we not negotiated at all.* In other words, the agreement you negotiate with this particular partner should be better than any other agreement you could have reasonably arrived at with any other negotiating partner. The agreement should also be better than if you had maintained the status quo rather than negotiating an agreement at all.

2. *Satisfy the interests of all parties to the negotiation.* We have mentioned before that it is imperative in a negotiation to dig through the initial positions of the parties to reveal their true intentions. You must all get to the heart of what each party really wants from the process. If you are able to successfully get to the root desire of each party to the negotiation, you will at least be in a position to satisfy those desires successfully.

3. *Select the best possible option to the negotiation.* This may sound obvious, but the point here goes back to the point we made earlier about searching for creative solutions. If you put your biases behind you and really look for options that go beyond what you first imagined possible, you will be able to find the best option to satisfy all your negotiating partners and yourself, too.

4. *Ensure everyone leaves the table happy.* A negotiation cannot be considered successful if one party's satisfaction comes at the expense of any of the other parties. A win-win outcome is only possible if you rely on objective standards that everyone agrees to beforehand and if you truly search for a creative solution. All parties to the negotiation should participate not only in arriving at the solution, but also at the ground rules for the negotiation in the first place.

5. *Get it in writing.* Any commitment made during a negotiation should be in writing and all parties at the table should agree that they understand the commitments before you move on in the process. At the end of the process, every detail of the agreement should

be recorded and all parties to the negotiation should sign off on the agreement, affirming that they understand and agree to every detail.

6. *Communicate.* All parties to the negotiation should feel comfortable communicating with each other. If this is not the case from the beginning, steps should be taken to establish these communication links before the negotiations begin. You cannot be successful at a negotiation if any party cannot communicate with the other parties.

7. *Remember that working relationships between the negotiating parties are better at the end of the process than at the beginning.* A successful negotiation process should build bridges, not burn them. If you are negotiating with a particular party now, there is a good chance you will have to negotiate with that same party again. Think how much easier it will be to negotiate the second time if the process worked well for you both the first time. A successful negotiation process should help build trust between the parties and should be an encouragement for the parties to work together again in the future.

Listening skills

Finally, possibly the most important skill a project manager needs to exercise during any negotiation process is listening. It is impossible to negotiate successfully with another party if you don't hear—and understand—what that other party is saying. It is difficult at best to persuade another party when you don't know the motivating factors behind that party. There are three major listening pitfalls for negotiators:

1. *Actively talking/passively listening.* A good negotiator must actively listen so that he or she will know what to say when they talk.

2. *Overpreparation for their next speaking opportunity.* Negotiators should avoid preparing for their next turn to speak while their negotiating partners are talking.

3. *Selective listening.* Negotiators should be open to listening to everything that is said, even if they don't want to hear it.

Listening skills can be divided into two groups, attentive skills and interactive skills. Attentive listening skills are used to better receive the true meaning of whatever your negotiating counterpart is trying to say. Interactive skills are used to verify with the sender that you understand his or her message. Attentive skills might include the following:

- *Be motivated to listen.* Typically, the person with the most information performs the best in a negotiation and what better way to gather information than to listen and comprehend what your negotiating partners are saying.

- *Ask questions.* Questions help you get to the root motivations behind your negotiating partner's positions.

- *Watch for body language.* Gestures, facial expressions, and tone of voice may convey more meaning than the actual words that are being spoken.

- *Let the other side speak first.* Again, you gain valuable knowledge and you have the opportunity to tailor your words to the needs of your negotiating partner.

- *Do not interrupt.* Interrupting someone who is speaking is rude and it also prevents you from gaining information that might turn out to be vitally important.

- *Avoid distractions.* You can avoid distractions by picking the environment for the negotiation carefully in the first place.

- *Record everything.* Don't trust your memory to accurately store all the information that comes from your negotiating partners.

- *Have a listening goal.* Listening for key words and phrases will help you respond more effectively when it is your turn to speak.

- *Keep your cool.* Don't overreact to what is said, and don't react to the speaker, only the words.

Interactive skills include the following three basic categories:

1. *Clarifying.* Ask questions to insure you heard what the speaker wanted you to hear. Ask the speaker to repeat or rephrase any part of his or her dialogue on which you aren't clear.

2. *Verifying.* Rephrase and repeat the speaker's words back to him or her. This gives the speaker an opportunity to restate any points that may not have come across clearly the first time around.

3. *Reflecting.* Reflecting involves showing empathy to the other speaker. You reflect the feelings and emotions of the speaker back to him or her to make him or her feel that you truly understand what they are saying and why they are saying it.

Project managers must be artful negotiators if they are to be successful. Negotiation for the project manager is not only limited to contract negotiation. Not only do project managers negotiate to secure agreements, they also negotiate to bring dissent to the surface which eventually may lead to creative solutions to complex problems. If you think about it for just a minute, the project manager's job requires a great deal of intuition about dealing with and persuading people. Most project managers spend more time negotiating than their counterparts in the contracting arena. Project managers must continually challenge the opinions of their project team members, their stakeholders, and their project sponsors as they search for the underlying motivations behind those opinions. Getting to these root motivations requires the project manager to utilize his or her negotiation skills to their fullest.

Reference

[1]   Ertl, D., *How to Build an Effective Negotiation Strategy,* Cambridge, MA: Vantage Partners LLC, 1997.

9

Conflict resolution

Introduction

Hand in hand with negotiation strategies goes conflict resolution. In fact, sometimes the two are inseparable. This chapter will look at why conflict exists, how to detect it in its early stages, and how to deal with it successfully so that it does not become a project killer.

Why conflict exists

One of the primary reasons conflict exists in organizations today is the rapid pace of change all organizations currently endure. The Industrial Revolution ushered in an era of organizations that were highly bureaucratic and roles were clearly defined. The past 30 years has seen an upheaval in organizational structure brought on by technological advances, new educational concepts, increased demand for leisure time, and societal concerns for the environment. These changes have led to disagreements over proper management structures and a lessening of clear definition of roles for workers.

Another cause for organizational conflict is the incongruity of organizational goals and employee goals. Organizational goals are dictated by top management, sometimes with no regard for individual employee goals. Today's business climate demands a more projectized organizational structure, thus leading to competition between functional and project managers in goal setting and competition between individual employees and the organization as a whole in which goals to pursue. "The conflict specifically associated with project management may be classified into two broad, partly overlapping categories: (a) conflict associated with change; and (b) conflict associated with the concentration of professionals of diverse disciplines in a more or less autonomous group effort which has limited life" [1].

Conflict is inevitable in organizations because organizations do not have unlimited resources with which to deal with the myriad desires of the individual employees. Therefore, conflict must be effectively managed within the organization, or even more scarce resources will be lost as a result of decreased efficiency and production from employees. The project environment is particularly conducive to conflict and thus, it is imperative that project managers become adept at dealing with it.

Conditions that lead to conflict

Conflict is generally defined as a clash between opposing elements or ideas. It can range from mild disagreement to an emotionally charged confrontation. Two basic, opposing views of conflict exist. The traditional view labels conflict as bad. In this view, conflict is caused by troublemakers and should be avoided at all costs. Managers who hold this view try to conceal conflict and will even deny that it exists. The contemporary view of conflict admits that conflict is inevitable. It is a natural result of change and, if properly managed, may be beneficial. How can an organization expect to generate new ideas if it is not willing to deal with a host of differing ideas coming together all at once? In order to manage conflict effectively and to use it for benefit, it is necessary to understand the conditions that lead to conflict [1]:

- *Ambiguous roles and responsibilities.* If roles and responsibilities are not clearly defined, individuals or groups may go beyond the boundaries of their defined tasks and come into conflict with other individuals or groups. Roles and responsibilities may also become ambiguous when individuals or groups report to two different managers at the same time, such as may happen in a matrix organization.

+ *Conflict of interest.* Two or more individuals or groups working toward inconsistent goals will lead to conflicts of interest. For example, a front-line worker wants to add lots of bells and whistles to a product so he or she may then tout their creativity, while the manager of the project wants to stick to a standard design so the project can be brought in on time and within budget.

+ *Barriers to communication.* Communication barriers come in all shapes and sizes, but regardless of the type, they inhibit understanding between parties and lead to conflict.

+ *Dependence.* One party depending on another for work inputs can lead to conflict if the supplier party performs in such a manner as to adversely impact the ability of the receiving party to perform their function. The receiving party may also come into conflict with the supplying party if they perform in such a manner as to cause the supplier to perform an inordinate amount of rework or resupply.

+ *Differentiation.* Differentiation refers to the way organizations divide complex tasks into subtasks that are handled by different groups. While this is a necessary method of getting the work done, it inevitably leads to conflict between groups.

+ *Association.* This goes hand in hand with differentiation. If organizations assign subtasks to different groups, at some point these groups must associate with one another to accomplish the larger task.

+ *Consensus.* A need for consensus may lead to conflict because rarely, if ever, do different work groups within an organization agree so completely on any one issue. Consensus calls for everyone to subordinate their wishes to the greater good of the organization.

+ *Behavior regulation.* In certain instances—areas of high security for example—an individual's behavior may have to be monitored and strictly regulated. Most individuals resist strict behavior regulation and thus will conflict with management over this issue.

+ *Prior conflicts.* Further conflict will certainly arise when previous concerns are not addressed sufficiently. Small issues become large issues when prior conflicts continue to simmer underneath the surface.

Project managers must learn to recognize these conditions and they must channel the conflict either to a resolution or to constructive purposes. For example, conflict can be used to stimulate a search for new and innovative ideas that are mutually acceptable to all parties involved. Conflict may also be used to enhance group cohesiveness by forcing the parties involved to explore the motivations of their adversaries and thus learn more about them. This will lead to new understandings between the parties and a greater sense of togetherness.

Dealing with conflict

There are five distinct methods for dealing with conflict and project managers must carefully choose the method most appropriate for any given situation in order to achieve positive results [2]:

1. *Smoothing.* Smoothing is deemphasizing differences and emphasizing commonalties.

2. *Withdrawal.* Withdrawal is retreating from potential disagreements.

3. *Compromising.* Compromising means considering various aspects of a situation, bargaining, and searching for mutually satisfying solutions.

4. *Forcing.* Forcing is simply exerting one particular viewpoint over all others.

5. *Problem solving.* Problem solving involves actually addressing a disagreement, defining the problem, developing alternative solutions, testing these alternatives, and selecting the most suitable one for the situation.

No one method will be a fit for every situation. Problem solving may not work if you are in a time crunch. Smoothing and withdrawal are simply delay tactics that don't resolve anything. Forcing is a rapid short-term solution, but may lead to long-term problems. Compromise provides a solution, but it's usually one that no one is particularly happy with.

In addition to these five distinct methods, there are five personal styles for handling conflict which may be related to the five methods [1]:

1. *Win-lose.* Win-lose involves a high concern for personal goals and a low concern for relationships;

2. *Yield-lose.* Yield-lose involves a low concern for personal goals and a low concern for relationships;

3. *Lose-leave.* Lose-leave involves a low concern for personal goals and a low concern for relationships;

4. *Compromise.* Compromise involves a moderate concern for personal goals and a moderate concern for relationships;

5. *Integrative.* The integrative style involves a high concern for personal goals and a high concern for relationships.

The win-lose person is a battler who seeks to meet his or her goals at any cost. The yield-lose person is a helper who places too much value on relationships to the expense of his or her own goals. The lose-leave person takes the view that conflict is hopeless, useless, and punishing. The compromising person will look to find a position where everyone gets something. The integrator seeks to satisfy his or her own goals and those of the other parties too. These last two styles typically are the most effective in drawing positive results from conflict in that they are focused on positive outcomes.

Now that we have overviewed conflict in general, let's take a look at conflict within projects and project organizations. We'll start with matrix organizations, as this seems to be the structure which contributes the most to organizational conflict. A matrix structure is one in which the project manager is in charge of his or her human resources for that project only, while the functional managers retain a great deal of control over the human resources that are borrowed from them for the project. Project personnel must take direction from two (or more) different people who may or may not have the same goals. Ambiguous jurisdictions, as mentioned earlier, lead to conflict. In addition to ambiguous jurisdictions, there are seven other sources of conflict unique to the project environment [3]:

1. *Project priorities.* The project manager, project participants, stakeholders, sponsors, and functional personnel may differ over the activities defined for the project or over the sequencing of the project activities.

2. *Administrative procedures.* The project manager and functional managers may conflict over administrative procedures specific to the project. For example, there may be a disagreement over how time devoted to project activities is recorded.

3. *Technical options.* Usually when technology is involved, a project team will want the latest and greatest technology, while the functional managers are less likely to introduce new hardware or software into the existing production structure. Budget issues often arise when technical options are discussed.

4. *Manpower.* Project personnel are often pulled from functional areas and conflict arises over who is chosen to work on the project and how much time these people may devote to a particular project.

5. *Cost.* Cost estimates must be made for the tasks on the project WBS. These estimates may come into conflict with existing budgetary numbers.

6. *Schedule.* Project schedules may run into conflict with existing functional schedules both from a manpower aspect and from a materials aspect. Project implementation will also have to be coordinated with existing production schedules.

7. *Personality.* Personality issues will always be a source of conflict in organizations, but the nature of project teams and a project organization puts a spotlight on these particular issues.

Project conflict may also be defined by the life cycle phase of the project. Listed below are the sources of conflict associated with each phase of the project life cycle in the order of which antecedent conditions causes the most conflict in that particular phase [3]:

◆ Concepual phase
   1. Project priorities;

2. Administration procedures;

3. Schedule;

4. Manpower;

5. Cost;

6. Technical options;

7. Personality.

- Planning phase:

  1. Project priorities;

  2. Schedule;

  3. Administration procedures;

  4. Technical options;

  5. Manpower;

  6. Cost;

  7. Personality.

- Implementation phase:

  1. Schedule;

  2. Technical options;

  3. Manpower;

  4. Project priorities;

  5. Administration procedures;

  6. Cost;

  7. Personality.

- Close-out phase:

  1. Schedule;

  2. Manpower;

  3. Personality;

  4. Project priorities;

  5. Cost;

  6. Technical options;

  7. Administration procedures.

Bases of power

It is obvious that the project manager must recognize all conditions that can lead to conflict and must learn to deal with conflict as it arises in the most constructive manner possible. The question now arises, How does he or she accomplish this task? Why should anyone listen to the project manager? He or she typically does not control the human resource purse strings, and while he or she can pass negative comments back to the functional manager, typically the functional manager is still in charge of the individual's performance review. The project manager must rely on the five bases of power that can be used to influence other people [4]:

1. *Legitimate power.* Legitimate power is formal authority based on a project manager's position in the organization and the willingness of project participants to recognize that position.

2. *Coercive power.* Coercive power is based on fear of reprisal if the subordinate does not perform in a manner satisfactory to the superior.

3. *Reward power.* Reward power is derived from the ability to use positive reinforcement to reward a desired behavior.

4. *Expert power.* Expert power comes from having unique knowledge pertinent to the particular situation at hand.

5. *Referent power.* Referent power derives from association with someone who is perceived as powerful (e.g., the boss's son, a project manager hand-picked by a VP).

A project manager may adopt any or all of these bases of power depending on his or her degree of control, position in the organization, knowledge and experience, and the organizational climate in which the project operates.

Conflict in organizations is unavoidable. A project structure within organizations just increases the likelihood for conflict. It is therefore necessary for a project manager to learn to deal with this inevitable conflict and to learn to channel this conflict down productive avenues. He or she must learn to recognize the precursors to conflict so that he or she can prepare to employ the proper leadership style with which to deal with the upcoming conflict. Finally, a project manager must learn to use all avenues of influence and power available to him or her to deal effectively with conflict on project teams.

References

[1] Filley, A. C., *Interpersonal Conflict Resolution*, Glenview, IL: Scott, Foresman and Company, 1975.

[2] Blake, R. R., and J. S. Mouton, *The Managerial Grid*, Houston, TX: Gulf Publishing Company, 1964.

[3] Thamhain, H. J., and D. L. Wilemon, "Conflict Management in Project Oriented Work Environments," *Proc. 6th Annual Seminar/Symposium*, Newtown, PA: Project Management Institute, 1974.

[4] Adams, J. R., S. C. Barndt, and M. D. Martin, *Managing by Project Management*, Dayton, OH: Universal Technology Corporation, 1979.

more personally, but help their subordinates accomplish more too. She has identified a number of indicators of a manager's power, which are as follows [2]:

1. Powerful managers intercede favorably on behalf of subordinates;

2. Powerful managers obtain desirable placement for talented subordinates;

3. Powerful managers are able to obtain finances beyond those budgeted;

4. Powerful managers influence corporate policy agendas;

5. Powerful managers have access to top decision-makers;

6. Powerful managers have early access to information about corporate policy shifts.

Power defined

We should at this time try to define what we mean by power. Traditionally, power is viewed as having control over others, but the changing business climate is redefining it as the ability to get things done. That changing climate encompasses the following trends:

- *Organizational flattening.* Layers of management are being eliminated in many organizations. Position in the organization is becoming less of a source of power. The ability to work across functional lines to accomplish goals has taken on more importance.

- *Information flow has been decentralized.* Information technology has made it possible for any worker in an organization to communicate with any other, regardless of position in the organization. Superiors cannot control information flow to or from their subordinates as they once could.

- *Projectization of organizations.* As traditional hierarchies break down and more functions are performed via project teams, the value of getting tasks accomplished takes on greater weight. Control over others is less of an issue on a project team than completing the project on time and within budget.

How does one go about gaining a base of power within an organization? Two factors seem to determine an individual's power: personal charisma and positional attributes. The influence each of these factors has will depend on the type of organization. It is a good idea for project managers to try to develop both their personal and positional attributes to their full potential, as neither factor by itself will sustain you indefinitely. Let's take a look now at the components of both personal charisma and positional attributes.

Personal charisma

Expertise is the first component of personal charisma. As technology advances and jobs in organizations demand the use of more and more technical skills, expertise in a particular skill set will become more important. Technology has also created the situation where a subordinate may actually possess more power than his or her superior. In the high-tech world a dichotomy emerges as individuals move up the corporate ladder. Some individuals choose to take the management route up the ladder while others choose a technical path. Expertise is absolutely essential to gathering a power base if you choose the technical path. The development of expertise in a particular area does not come without its drawbacks. Technical specialists are often typecast as just technical specialists and are not given consideration for movement into other areas of the organization.

Personal attractiveness is the next component of personal charisma. This attribute may also be referred to as likability. This is not to say that everyone who is well liked at the office has a great deal of power. It is also not to suggest that you should try to be close friends with all your coworkers and business associates. What we can say is that the same characteristics that place a person high on the likability scale will make this person more attractive to coworkers and will thus assist in boosting the individual's personal power. Much research over the years supports the fact that likable people tend to get the benefit of the doubt more often, tend to be more persuasive, and tend to be viewed as more trustworthy than their less likable counterparts.

Effort is the third component we will discuss. Old adages, such as put your nose to the grindstone and put your shoulder to the wheel, come to mind. While expertise and personal attractiveness are important, there is no substitute for hard work in building a base of power. Extra effort typically leads to increased expertise and increased understanding of organizational processes. Individuals who put out extra effort on their tasks are usually viewed as contributors to the common good of the overall group, and thus, they are able to accrue power through their efforts.

The final component we will discuss is legitimacy. Every organization has a set of norms, values, and expected behaviors that management attempts to instill in all employees. Those employees who conform to these expectations gain legitimacy in the eyes of management and are typically more likely to be advanced up the organizational ladder. In a unique paradox, once these organizational members gain power through legitimacy, they find themselves in a better position to challenge the norms and values that got them the power in the first place.

Positional attributes

One positional attribute that may lead to power for an individual is his or her placement in the communication flow of the organization. No one person in an organization has all the information or all the knowledge necessary to do any particular job. A network of individuals is necessary to accomplish organizational tasks. The more complete the network you can build, the more chance you have of being successful in the organization. In addition, the more central and more critical you can make your position in this network, the more powerful your position. Your communication network should be both horizontal and vertical within the organization. Be willing to cut across functional and departmental lines to expand your network. Look for opportunities to make yourself valuable to both your immediate workgroup and to other workgroups outside your area. Ask yourself if you were gone for a week, what sort of problems would it cause for the organization? The idea is to make your position both central within the organization and critical to the organization.

Another power-enhancing attribute for individuals in organizations is flexibility. Individuals with the ability to improvise or make decisions without approval from superiors hold more power. One way to determine the amount of flexibility an individual possesses is to examine the reward system in place for that individual. If an individual is rewarded for reliable, predictable performance, then they have less flexibility than an individual who is rewarded for innovation and creative thinking. Individuals seeking power should look to find positions that are rewarded in the latter manner.

Visibility is also a key element in obtaining power within an organization. Excellent work is essential, but someone must notice in order for you to gain power from the quality of your work. One measure of visibility is the number of people with whom you interact on a daily basis, and the position in the organization these people hold. It is always a good idea when looking to build your power base that you make contact with as many senior-level executives as possible. There is no substitute for face-to-face contact when trying to increase your visibility in an organization. Take every advantage to participate in projects whose results will be presented to top management, especially if you can make at least part of the final presentation.

The final attribute we will discuss is relevance. The position an individual holds must be relevant to the survival of the organization. For example, a software engineer is more closely tied to the survival of a software development firm than to the survival of an electric utility. An individual in pursuit of power should assess his or her skill set and then determine to what type of organization this skill set is most critical.

Let's switch gears somewhat at this point and discuss a different but related topic. We have talked at length about power and how to obtain it. However, power by itself is useless. Gaining power for power's sake typically only leads an individual to rejection by an organization. The real benefit of acquiring power is in having the ability to transform that power into influence. It is really influence that allows individuals to accomplish tasks and go above and beyond objectives.

We will discuss three strategies for exerting influence on others: retribution, reciprocity, and reason. Each strategy has its own manner of implementation and each is most appropriate under differing circumstances.

Retribution: Retribution involves the use of a threat by a superior on a subordinate based on the superior's formal authority. The threat can be explicit (you'll lose your job if you don't do as I say), or implicit (a subordinate is consistently ignored in meetings). In either case, the subordinate understands that he or she will have unpleasant consequences with which to deal if he or she does not adhere to the superior's expected behavior. Retribution may also come in the form of corporate intimidation. I have worked at organizations in the past where considerable arm-twisting was applied to achieve desired results—for example, attainment of a certain contribution level to a charity. I have also seen fear instilled in employees to achieve certain levels of participation in political-action committees with the warning that we might lose our jobs if we don't get the right politicians elected who will vote our way.

Reciprocity: Reciprocity rests of the notion of the win-win situation. The self-interest of both parties involved in an issue is served in a reciprocal situation. Reciprocity can come in simple forms, such as asking employees to work overtime in exchange for comp-time, or it can be more complex, such as the case of winning support from company leaders for a new project in exchange for incorporating their wish lists into the project plan. Reciprocity still relies on a supervisor's ability to control outcomes valued by subordinates, but it does so in a way that does not discourage or frighten the subordinate, thus leading to a healthier relationship between subordinate and superior.

Reason: This approach is used to acquire compliance based on the merits of the request itself regardless of the relationship between the supervisor and the subordinate. The requestor makes a statement of fact that appeals to the other person's sense of values and goals, then makes his or her request. Keep in mind that the request should be explicit and direct so as not to appear manipulative.

The three strategies are summarized in Table 10.1 [3]. It is not only important to understand the techniques you have available to persuade others to do your will, it is equally important to learn how to deal with these strategies when others are using them on, with, or against you. We will discuss three countermeasures to each strategy beginning, in each case, with the method to use initially and then proceed on to methods two and three.

Antiretribution strategies

1. *Build your power base early to offset perceived inequities in power between you and your superior.* The more inequity the superior perceives, the more he or she will rely on retribution as a strategy. Try to interject other means of accomplishing your superior's wishes into your discussions.

TABLE 10.1　Three Strategies for Exerting Influence on Others

STRATEGY	WHEN USED	ADVANTAGES	DISADVANTAGES	OTHER ISSUES
Retribution	1. Requestor has power. 2. Quality not of high importance. 3. Request is time constrained. 4. Request is specific. 5. Resistance is likely.	Response is quick and direct.	Harms creativity and commitment.	May be a violation of ethics and employee rights.
Reciprocity	1. Parties depend on each other. 2. Each party has a resource of value to the other. 3. Time constraints are less critical. 4. Request is specific. 5. Each party views the other as trustworthy.	Resentment between parties is of a low incidence.	Encourages employees to view every-thing as up for negotiation and encourages employees to always expect some reward for their coop-eration.	Possible manipulation and chance for unfulfilled expectations.
Reason	1. Time is available for discussion. 2. Requestor and requestee share values and goals 3. Mutal respect between parties exists.	Surveillance of trans-action is necessary as it goes forward.	Requires common trust, common respect, and common values and goals.	Hard to obtain agreement based on reason, but easy to derail.

2. *Confront the exploitation head-on.* Describe how the retributional tactics impact you in a firm but reasonable manner. If necessary, outline steps you intend to take if the tactics don't stop.

3. *Fight fire with fire.* This involves directly resisting the request, whether through reduced performance or simply refusing the

request. This is a last option response as it is almost sure to invoke adverse consequences.

Antireciprocity strategies

1. *Examine the motives of the requestor.* Question what the requestor expects in return for his or her offer. Decline the offer if you believe you will be adversely affected.

2. *Confront manipulators directly.* Point out what you perceive as manipulative tactics and then make a counteroffer that is more favorable to you than the original offer.

3. *Refuse to bargain with manipulators.* Back away from the negotiation to make sure you would or would not accept the offer under circumstances where no duress was present.

Antireason strategies

1. *Place the request in the proper light from your perspective.* Someone else's must-have may only be a nice-to-have for you. Even a reasonable request may not be a good deal for you due to resource constraints.

2. *Stand up for your personal rights.* Remember that you still have the right to refuse a request, even if it is perfectly reasonable. Don't always subordinate your wishes to those of others.

3. *Refuse to comply.* If the requestor continues to persist, be direct and simply refuse. Some people only understand this direct approach, and the direct approach may head off future requests that similarly place undue demands on your resources.

Finally in this chapter we will examine the issue of wielding power with superiors within your organization. In order to rise above the din of issues that constantly assault upper management, you must learn how to sell your issues and you must make your superiors see how you can benefit him or her above and beyond other subordinates in the organization. Listed below are important points to keep in mind as you attempt to exert influence upward within your organization.

- Selling your ideas:
  - *Congruence.* Only sell issues congruent with your role in the organization.
  - *Credibility.* Be honest. Don't be self-serving. Be sincere.
  - *Communication.* Broaden your network of organizational communication.
  - *Compatibility.* Sell issues that mesh with company goals and objectives.
  - *Solvability.* Show that the issue can be brought to fruition.
  - *Payoff.* Demonstrate clearly the payoff for the company. The more immediate the better.
  - *Expertise.* Advance issues where the relevant expertise currently resides within the organization.
  - *Responsibility.* Tie top management responsibility to the organization to the issue you are advancing.
  - *Presentation.* Present your issues succinctly, clearly, and positively.
  - *Bundling.* Show how your issue can be tied to other current issues that are receiving priority.
  - *Coalitions.* Enlist the help of others who have influence with top management to help sell your issue.
  - *Visibility.* Select the most visible forum possible in which to sell your issue.

- Benefiting the boss:
  - *Solve problems.* Find out the most pressing problems for your boss and develop solutions to them.
  - *Understanding.* Develop an understanding for the challenges that face your boss every day.
  - *Diagnosing.* Examine your boss's strengths and weaknesses. Learn where you can fill in the gaps.
  - *Self-awareness.* Assess your own strengths and weaknesses. Use your strengths to overcome your boss's weaknesses.

- *Communicate.* Keep your achievements front and center with your boss.

- *Build trust.* Do your work well and keep all your cards on the table with your boss. Avoid deception at all costs.

- *Protect.* Shield your boss from minutiae and help free his or her time for what he or she considers most important.

- *Listen.* Listen to as many different perspectives on the issues as possible so that you might be as widely informed as possible when discussing these issues with your boss.

- *Speed.* Complete tasks accurately, but with speed. Continue to exceed your boss's expectations on completion times.

- *Be creative.* Be an idea generator for your boss.

Conclusion

Power is not given, it is earned. Power can originate from a variety of sources and it is to your benefit to develop as many of these sources as possible. Once power is gained, it must be translated into influence if it is to benefit the one who holds it. Influence can be directed downward within the organization, or upward as is the case when you are attempting to sell ideas to upper management. Power and influence must not be abused or you will foster mistrust and resentment among your peers and with your superiors. It is the proper use of power and influence that separates the truly successful individuals in an organization from those who flounder or burn out.

References

[1]   Bennis, W., and B. Nanus, *Leaders*, New York: Harper and Row, 1985.

[2]   Kanter, R., "Power Failures in Management Circuits," Cambridge, MA: *Harvard Business Review*, 1979.

[3]   Whetton, D., and K. Cameron, *Developing Management Skills*, Reading, MA: Addison-Wesley, 1998.

11

Empowering others

Introduction

We focused, in the last chapter, on gaining power and influence for yourself. This chapter will look at how we can distribute that power to others around us, thus increasing the productivity and job satisfaction levels of those we empower.

Empowerment

Empowerment requires that we loosen controls and let our employees or team members set goals, make decisions, and achieve accomplishments for themselves. This is not a natural process for most people as we tend to want to be in control and tell others what to do. Once a project leader recognizes, however, that empowering team members actually increases his or her own power base, it becomes a more palatable concept.

A natural tendency in an uncertain climate is for people to rely on the tried and true habits of the past, regardless of whether these habits are effective in the current climate. People become more conservative, less willing to make decisions, and less communicative. The problem is that in unsettled times, just the opposite characteristics are needed. Employees need to be flexible, autonomous, and open to two-way communication. Table 11.1 [1] illustrates a number of survival attributes employees may exhibit in an unstable business climate. A project manager must be on the lookout for these attributes and must learn to counteract these behaviors as soon as they appear.

The question for project managers then becomes how do we avoid these reactive behaviors given the current pace of change in the business world today. Empowerment is one key to overcoming these behaviors. This means that you enable your people to take action, to make decisions, to have control over their situations. When you empower employees you help them have intrinsic satisfaction from their work situation, and as we have discussed in a previous chapter, intrinsic satisfaction keeps people happier than any external motivator you can offer.

TABLE 11.1   Survival attributes of employees in an unstable business

Behavior	Comments
Short-term perspective	Long-term planning is abandoned and employees begin working in crisis mode.
Scapegoating leaders	Organization leadership is questioned and criticized by employees.
Loss of loyalty	Commitment to the organization erodes. Employees become defensive.
Loss of teamwork	Coordination among employees deteriorates.
Loss of communication	Information is tightly held. Bad news is not communicated.
Increased conflict	Employees become self-centered to the detriment of the organization as a whole.
Loss of trust	Distrust of leadership predominates among employees.
Increased politicization	Employee body splinters into special interest groups.
Decreased morale	The mood of the organization as a whole turns dark. Employees stop enjoying their work.
Loss of innovation	Creativity is stifled as there is a low tolerance for error.
Resistance to change	Employees become more conservative and self-protective.
Centralization	Power is consolidated at the top of the organization.

What sort of roadblocks should a manager expect to experience when attempting to empower his or her team members? There are three major groups of inhibitors to empowerment [2]:

1. *Inferior view of subordinates.* Some managers refuse to let go of the reins because they do not believe their team members are competent enough to make good decisions. The fact is, employees are often capable of making appropriate decisions, but not necessarily in the manner that the manager makes those decisions. It's the my-way-or-the-highway syndrome where the manager can't deal with the fact that there's more than one way to attack a problem and get it solved.

2. *Insecurity of management.* Another reason managers refuse to empower employees is that they fear they will lose some of their own recognition and authority if they share decision-making duties with subordinates on their team. Sometimes the manager may feel that a subordinate may actually surpass him or her on the corporate ladder if that subordinate is given too much authority.

3. *Fear of losing control.* Finally, some managers just don't have the confidence to give up total control of every activity of their work group. They hold to the belief that employees must be directed or else confusion will reign.

These three inhibitors all are based on what I like to call an assembly-line theory of management. Managers who hold to the above principles believe employees are just cogs in the machine that must be managed to the nth degree. They believe in hierarchical management structures and a clearly defined chain of command. We have seen in previous chapters that this type of thinking and management style is counterproductive in a highly technical, highly volatile business climate. Not only that, this type of thinking can have long-lasting psychological impacts on employees by diminishing their self-worth and decreasing their motivation to perform in other aspects of their life outside work. Empowerment doesn't just make good business sense, it is ethically the right thing to do for your employees.

In order for managers to successfully empower others, they must instill five qualities in their subordinates [3]:

1. *Self-efficacy.* This quality involves (a) a belief that you have the ability to perform a certain task, (b) a belief you can put out the effort necessary to perform the task, and (c) a belief that you can overcome all obstacles in accomplishing the task. It boils down to employees possessing the capability and the confidence to perform any task put in front of them.

2. *Self-determination.* Empowered employees feel they can involve themselves in tasks because they choose to do so, rather than because they are made to do so. As a result of these feelings of control, empowered employees take ownership and responsibility for their actions, thus increasing both the quality of, and accountability for, the work that is done.

3. *Personal control over outcomes.* Once you empower individuals, they develop a feeling that external job factors can be controlled, rather than a feeling of external factors controlling them. Employees need to feel that what they do produces an effect and that they are in control of that effect in order for them to feel empowered. Personal control over outcomes has also been linked to emotional health and well-being. People who live in situations where they do not have this control often exhibit psychopathic behavior.

4. *Personal significance.* Another aspect of empowerment is helping employees have a feeling that their work means something. Empowered individuals invest part of themselves in their activities. They attach their own value systems to their work and develop a sense of connectivity with their outputs. People who engage in meaningful work activities are energetic and enthusiastic about their work, and this energy and enthusiasm is infectious within the work group. Empowered individuals have this feeling of meaningfulness about their work and are more committed to their work activities than those employees who are not empowered.

5. *Trust.* Empowered employees have a sense of confidence in both their superiors and subordinates. They are secure in themselves and the work they are performing. They are also more open, honest, and willing to learn new things. Trusting employees work

better in team situations due to their ability to communicate openly and directly.

Now that we have outlined the five key attributes of empowerment, we must ask ourselves how we instill these attributes into our team members. Traditional bureaucracies tend to instill dependence and submission to authorities in their employees. Project teams, particularly in high-tech environments of today, must instill just the opposite attributes. Namely, project team members must be empowered. Let's look at nine key elements managers can use to empower team members [4]:

1. *Clear goals, vision, and mission.* Team members must have an idea of where it is they are to go if they are to make decisions about how to get there. The most effective way to articulate goals, visions, and missions is by way of graphics and relations to real life. Engage both the left and right brain of your team members.

2. *Personal mastery experiences.* It has been demonstrated that the single best way to empower employees is to help them master some challenge. Suitable tasks must be provided to the individual along with the tools to handle the task. Also, celebrate even the smallest victory. There is an old Native American bit of wisdom that says the longest journey begins with a single step. Large problems are difficult to conquer all at once, but when you break them down into their component parts, they don't seem quite as daunting.

3. *Modeling.* People must believe tasks are doable. One way to instill this belief is to provide models of others who have conquered a particular task. Managers can be role models in some cases, but often it is better to put forth peer-group role models and establish mentors within teams. In this way the manager can be sure the modeled behavior will be displayed to the fullest and will be available to the other team members when they need it to be.

4. *Social and emotional support.* Social and emotional support includes praise, encouragement, and approval. Empowered employees know their manager will go to bat for them if needed. Empowered employees also know that their achievements will be recognized adequately.

5. *Positive emotional arousal.* Positive emotional arousal does not simply mean throwing parties to celebrate employee accomplishment. Positive emotions are aroused in employees when they are given work to do that is closely connected to values that they hold dear. People will fight and die for a cause they truly believe in, whereas sometimes they won't even work for money to accomplish a task that they do not believe in.

6. *Information exchange.* Managers must be a free-flowing conduit of information between the outside environment and their project teams. The old saying, knowledge is power, applies here. The more information team members have, the better able they will feel to make decisions and take actions on their own. The worst situation a manager can be in, from an information-sharing point of view, is one in which employees believe the manager is withholding information. Human nature tends to make us invent situations that are typically far worse than the truth. It is much better for a manager to distribute both good and bad information as soon as possible. It engenders trust between the manager and the team members and fosters higher morale in the long run.

7. *Resources.* Obviously every employee can't be given carte blanche to acquire all the resources he or she would like to have to perform his or her job function. It is up to the manager to make sure that, at a minimum, each team member has the resources to accomplish what is expected of him or her. Resources don't just include physical equipment, they also include time, space, communication channels, and access to interpersonal networks.

8. *The big picture.* Another element of empowerment comes from allowing workers to tie the efforts they put forth with the final outcome of whatever product or service they are providing. In other words, managers should provide the means for their team members to see the big picture. In addition, employees are typically more satisfied when they are able to manage a process from end to end rather than just performing individual tasks within the process. Again, they are able to obtain the big picture view rather than just a narrow view of a piece of the total effort.

9. *Confidence building.* Employees must trust their management and they must trust one another. It is essential that all members of a team or an organization have confidence that everyone else will be straightforward and honest. Managers must be reliable, fair, caring, open, and competent if they expect team members to work hard and accomplish great things. If any of these qualities is lacking, it will be very difficult for the team to have confidence in its management.

By way of a review, Table 11.2 cross-references the five dimensions of empowerment with the nine methods for fostering empowerment among employees.

TABLE 11.2   Methods for Fostering Empowerment

DIMENSION	METHODS OF EMPOWERMENT
Self-efficacy	Personal mastery experiences
	Modeling
	Information exchange
Self-determination	Clear goals/vision/mission
	Information exchange
	Resources
	The big picture
Personal control over outcomes	Clear goals/vision/mission
	Personal mastery experiences
	Resources
	The big picture
Personal significance	Clear goals/vision/mission
	Social and emotional support
	Positive emotional arousal
Trust	Modeling
	Social and emotional support
	Confidence building

Delegation

Projects contain a myriad of tasks that require coordination among a number of people. One role of the project manager is to assign these tasks to the appropriate personnel and make sure these tasks are completed. The project manager does this by delegating work out to team members. Thinking back on our discussion of empowerment so far, you can see that delegation dovetails in here nicely. A project manager must empower his or her team members when he or she delegates work to them so that the task can be accomplished within the parameters of the project. Empowered delegation can be used as a tool to help team members experience personal mastery over certain items. It builds trust between the team members and the project manager. Finally, empowered delegation is a time-management tool for the project manager. By delegating tasks, and developing the project team to handle even more tasks in the process, the project manager is freed up to handle broader issues that relate to the project such as communication with stakeholders, resource tracking, and networking with influential project supporters. In order for a project manager to successfully empower his or her team members and to successfully delegate tasks to them, he or she must decide when to delegate tasks and to whom they should delegate those tasks. The project manager should ask five basic questions to determine when to delegate [5]:

1. *Do team members have the required information, tools, and expertise to perform the task?* Team members may not always have the knowledge or resources to perform certain project tasks, so the project manager must make sure he or she delegates tasks that are appropriate for the team members' skill levels. On the other hand, sometimes team members are actually more qualified than the project manager to perform certain tasks as they are often closer to the end clients and the day-to-day workings of the organization.

2. *How committed are team members to the project?* Participation in project decision making will increase commitment of team members to the final outcome. In addition, it allows team members to have more overall knowledge of the project workings, thus increasing their feelings of empowerment.

3. *Will the delegated tasks serve to expand the team members' capabilities?* Project team members will view delegation negatively if the

only tasks ever passed out to them are the ones no one else wants to perform. Team members will be much more receptive to delegation, particularly of tough tasks, if they see the opportunity to expand their knowledge base or skill level.

4. *Are values and perspectives common between management and team members?* The best way to create common perspectives between all parties to a project is to explain why tasks need to be accomplished. Don't fall into the trap of only sharing the what and when elements of a task without explaining why. This helps team members see the project from a broader perspective—the big picture.

5. *Is time available to delegate a particular task?* As with every other aspect of project management, time spent up front saves time later on. Spend the time to prepare team members for delegation of tasks so that you can save time later by being able to delegate to them.

Once the decision has been made to delegate tasks and the decision has been made to whom tasks will be delegated, a project manager should follow these 10 principles to ensure successful empowered delegation [1]:

1. *Have a clear goal at the outset.* The project manager must articulate a clear goal for a delegated task and must link the task with something of value to the team member if the project manager expects that team member to take the task and run with it.

2. *Delegate completely or not at all.* Specify clear boundaries for the task. Make sure the team member understands the organizational rules, time and budget constraints, and quality expectations of the results of the task before he or she ever begins working on it.

3. *Allow team members to participate in the delegation process.* Rather than just passing out tasks, give team members the opportunity to request them first. Project managers are not always at liberty to give team members free rein in selecting tasks, but whenever possible, give team members as much freedom as possible to decide which tasks to perform and under what time constraints the tasks will be performed.

4. *Provide the same amount of authority as responsibility.* Team members must have enough authority to make decisions in performing a task as they have responsibility for the outcome of the task. Often team members are expected to perform tasks, and then their hands are tied because they have to appeal to the project manager for the authority to get the job done.

5. *Remain within the organizational structure.* Delegate to the lowest level possible at which a task can be performed. This is the level where the most knowledge will be found about day-to-day operations. Also, keep the communication links open and follow whatever chain of command is in place in the organization. Circumventing organizational structure can lead to disempowering those employees who are bypassed.

6. *Support your team members.* Once a task is delegated, you must continue to supply pertinent information to the delegee. Provide the resources necessary to perform the task, within the boundaries of the project or organization. Praise team members publicly, but criticize them in private. Nothing tears an employee down quicker then to chastise him or her in front of his or her peers.

7. *Don't micromanage.* Define an acceptable level of performance on a task, outline the exact results expected from the task, then judge the team member on those metrics. Certainly all means do not justify the ends, but in order to empower employees, you must provide them the latitude to accomplish tasks in the manner that works best for them.

8. *Be consistent with delegation.* Don't wait until you have too many tasks to do yourself before you pass out tasks to team members. Delegate from the beginning, delegate both exciting tasks and non-exciting tasks, and let your team members know they are needed for their skills, not just to act as a warm body or an overflow pool of workers.

9. *Do not allow for redelegation.* Once a task has been delegated to a team member, that member should be responsible for seeing it through. As a project manager, you cannot get in the habit of accepting tasks back from your team members. Require them to

present solutions along with their problems when they come to you for direction. Do not allow them to become dependent on you to solve their problems. This defeats the whole process of empowerment.

10. *Define clear rewards and consequences.* Team members should understand from the start what the rewards will be for completing a task successfully and what the consequences will be if they are unsuccessful.

Conclusion

The high-tech business climate is counterproductive to empowering employees, so it is even more important now than ever that project managers learn the techniques for developing their team members to take responsibility and aspire to great levels of achievement. As is the case in other areas of project management, the key to empowerment and empowered delegation is communication. You must articulate clear direction, define clear boundaries, and use objective metrics by which to judge team member performance. By doing this, you engender trust in your team members and enhance their willingness to take responsibility and be accountable for their performance. In the end, empowered employees are healthier, happier, and more productive.

References

[1]   Whetton, D., and K. Cameron, *Developing Management Skills*, Reading, MA: Addison-Wesley, 1998.

[2]   Byham, W., *Zapp! The Lightning of Empowerment*, New York: Harmony Books, 1988.

[3]   Spreitzer, G., *When Organizations Dare: The Dynamics of Individual Empowerment in the Workplace*, Ann Arbor, MI: University of Michigan Press, 1992.

[4]   Bandura, A., *Social Foundations of Thought and Action: A Social Cognitive Theory*, Englewood Cliffs, NJ: Prentice-Hall, 1986.

[5]   Vroom, V., and P. Yetton, *Leadership and Decision Making*, Pittsburgh, PA: University of Pittsburgh Press, 1973.

12

Problem solving

Introduction

The last three chapters of this book deal with miscellaneous topics of interest to project managers and team members. These techniques are important to project teams if they are to function efficiently and become high performing. We'll start with problem solving. Solving problems is so intertwined into our daily work lives that most of us probably never give it a second thought. As project managers, we probably don't have time to think about problem solving; we just do it. This is particularly true in the high-tech world where technology and the business climate change so quickly that we are continually faced with new situations that require new solutions. Given the number of problems we face every day and the number of solutions we are expected to generate, it seems quite likely that some of those solutions may be less than optimal. We may aim for the first workable solution, instead of the best solution. Sometimes our deadlines may be so tight that we just grab the first available solution. Maybe it would behoove us to take a moment to think about the manner in which we solve problems. Rather than just developing some solution in the given amount of time, maybe we should try to find a way or a methodology by which we could develop the best solution in the given amount of time. This chapter will look at both structured and unstructured methods of problem solving and when each method is appropriate to use.

Structured problem solving

The first step in a structured problem-solving methodology is to actually define the problem. If our goal is to develop an optimal solution, we must solve the correct problem. Often we find ourselves applying band-aids to situations because we do not delve deeply enough into the situation to find the root cause of the problem. In order to define the correct problem we must first look for the factual information surrounding the situation. We must disregard what we suppose to be true or what our own predispositions tell us must be true. We must operate strictly in the factual world. We must gather as much data as we can on the problem from as many sources as we can. We must define why this situation is a problem in the first place. What pain is this situation causing and who is suffering that pain? Finally we must state the problem in clear, concise terms that are understandable by all parties involved.

Once a problem has been clearly stated, all possible solutions should be collected and examined, regardless of how improbable some of the solutions may be. Alternatives should not be evaluated at this point. The thrust of this step of the process is idea generation. Evaluation tends to squash idea generation at this stage. You should strive for a wide range of ideas from a wide range of people. Diversity of ideas and opinions enhance this step of the process. You may find that you can combine some alternatives to build a stronger alternative. You should also be cautious to stick to the problem at hand and not be distracted by alternatives that may address other problems.

Once your idea generators have been exhausted, you can move on to alternative evaluation. You should set your standards high in this step so as to look for the optimal solution, not just a satisfactory one. Each alternative should be given equal consideration with both the pros and cons of each alternative considered. The alternative selected must also be consistent with project or corporate norms at the time. For example, you typically do not have access to unlimited resources, so a solution that calls for more resources than you can access is probably out of the question regardless of how good it is.

The final step of the process is to implement the solution that has been chosen. As with project implementations, you should keep the lines of communication open with all parties involved with the implementation, you should maintain feedback loops with all these parties, and you should have an ongoing process of monitoring in place to make sure the solution actually solves the problem as you expected it to.

As you may have noticed already, this structured methodology makes some assumptions about the problems you are facing. First, this process only works for problems that are fairly straightforward. Alternatives that can be defined must be readily available, and metrics must exist against which to measure the alternatives. Ambiguity is not a friend of this process. Table 12.1 summarizes the steps to structured problem solving and their associated constraints [1].

Free-form problem solving

How do we attack problems for which there is no obvious solution, or for which no obvious alternatives exist even? A rational, structured approach is not appropriate in these situations. We must use our imaginations and our creativity to solve such situations. The problem is, the older we get and

TABLE 12.1   Steps and Constraints of Structured Problem Solving

STEP	CONSTRAINT
Definition	Consensus problem definition is difficult to achieve.
	Existing solutions may influence problem definition.
Generation of alternatives	Hard not to evaluate alternatives as they are proposed.
	Past success may influence alternatives proposed.
Selection of alternative	Alternatives may not be detailed enough for evaluation.
	Gathering information on alternatives may be costly.
	Hard to keep searching through alternatives to find an optimal solution as opposed to an acceptable solution.
Implementation	Acceptance for the solution may be hard to obtain.
	Resistance to change is common.
	Resources may not be available for proper implementation.

the more experience we gain, the less creativity we bring to bear on an issue. We become trained to look for the right answer and believe all our former experiences will provide us the data we need to analyze the situation and arrive at this one right answer. We create mental blocks for ourselves, blocks that prevent us from seeing the uniqueness of the situation. These blocks prevent us from seeing that maybe our past experiences have nothing at all to do with the current problem. In these situations we need to take a lesson from young children. They have no predisposed notions of how things ought to be, so they are able to examine new situations in an unfettered way that most adults cannot. Let's examine four types of mental blocks to free-form problem solving that adults develop over time [1]:

1. *Constancy.* Most humans are driven to achieve consistency in their lives. Once they find the solution to a problem, they will try to use that solution to other similar problems in the future, thus blocking the chance to find creative solutions to these problems. One way we fall into the constancy trap is by refusing to view problems from multiple angles. We must be open to various alternatives to a problem and we must also be open to various statements of a problem. Sometimes solutions become more apparent if a problem is just restated. Another way to fall victim to constancy is to restrict oneself to only thinking and communicating verbally. We must learn to rely on all our senses to creatively attack problems. Sometimes we use mathematical symbols to state problems. Sometimes we draw pictures. Sometimes a certain texture, or smell, or taste will trigger our thought processes. Fighting against constancy means fighting against our natural human instincts, but it is necessary if we are to find creative problem solutions.

2. *Commitment.* Commitment follows close on the heels of constancy in blocking our creative thinking processes. Once we become committed to an idea (or solution), we tend to stick with it, regardless of where that line of thinking leads us. We tend to try to group problems that are similar and then apply the same reasoning to all these problems in arriving at a solution. We are also influenced by past history and past research that tells us certain things are so. We often take this data as fact and fail to question it, thus preventing us from finding solutions to problems that may fall

outside the research of the past. Another commitment we make in our thinking is that all problems have to have unique solutions. Truly creative thinkers are able to see commonalties among seemingly unrelated events to arrive at solutions to seemingly unsolvable problems.

3. *Artificial constraints.* This block to creativity comes in two flavors. Sometimes we place constraints on a problem that do not exist, they are just drawn from our assumptions based on past history. We eliminate alternatives when we do not need to. The other extreme of this block is that sometimes we must constrain a problem sufficiently to find solutions to it. In other words, we must break large problems down into manageable pieces and we must filter out extraneous data that has nothing to do with the solution to the problem. This actually ties back to the concept from structured problem solving that teaches us we must clearly define the problem we seek to solve before we can begin to consider alternatives.

4. *Mental laziness.* Maybe the title of this topic is somewhat harsh, but it is meant to grab your attention and get you to focus on how slack our society has become when it comes to thinking. Our lives at work and at home have become so cluttered with activities, that very few of us take time to just think. When is the last time you found a quiet place to kick back and let your mind concentrate on possible solutions to some issue you were facing? We are so geared in to action and quick resolution that we look down upon those who would spend a few extra minutes thinking through an issue. In addition, most of us are too proud to ask probing questions for fear that we might appear less intelligent. In reality, the truly intelligent among us are the very ones who are not intimidated to ask questions, to probe, and to think before they act. If we are to become creative problem solvers, we must learn to utilize our whole brain, and we must be willing to expend some brain power before we choose to act on a solution to a problem.

Now that we recognize some of the impediments to free-form problem solving, let's look at ways that we can increase our ability to think creatively and solve these types of problems. First, let's examine the four stages of free-form problem solving [2]:

1. *Preparation.* This stage includes the activities of data gathering, problem definition, alternative generation, and alternative review. The preparation stage is the place where we can have the most conscious impact on the creative problem-solving process.

2. *Incubation.* This stage involves digesting the outputs of the preparation stage and combining them in myriad ways. This ties back to the discussion above on mental laziness. It is here that we need to sit back and just think.

3. *Illumination.* This is the culmination of the incubation period. An alternative is selected and articulated.

4. *Verification.* Verification involves validating the output of the illumination stage against known metrics to judge reasonableness and possible acceptance.

Since we can have the greatest direct influence on the preparation stage of the process, let's consider for a moment methods that will help us improve our problem-definition process. One way is to try to relate the unknowns in a problem to something familiar. Lest you think I'm contradicting the tenets of creative problem solving, let me say that I am not proposing to base solutions on the familiar, just that referring to the familiar may help frame the current problem so that you can move on to the next steps in the process. The use of analogies can be helpful in defining complex issues in simple ways. Another method to improve problem definition is to force yourself to generate multiple problem definitions. Once you have defined a problem, stop and see if you can define it again differently. Continue to ask yourself if there's more there to the issue. Consider different perspectives and try to think of how other people might frame this problem given the same set of circumstances. Finally, look at problems from the complete opposite viewpoint that you initially take. You may find that the problem you define the second time around is much different from the first and potentially much simpler to solve.

Another way to improve the problem-definition process is by taking steps to generate more alternative solutions to the problem. The most popular and most well known of these techniques is brainstorming. We've discussed brainstorming at length previously, and I'm sure you're all quite familiar with it. The key to brainstorming is to push beyond the initial rush of ideas so that you can really stretch for some creative alternatives. A second technique is to break a problem down into its component parts and concentrate on those individually. Sometimes this will lead to a longer list of alternatives than if you strictly look at the whole.

Just as we talked about empowering people and the methods for accomplishing that, so we should try to cultivate creativity and free-form thinking among coworkers and team members. Creativity in a vacuum is of little use, but a group or team of creative problem solvers can be just the seed to spread this type of thinking throughout an entire organization. How do we foster this type of thinking? Let's look at four possible ways to incubate creative thought:

1. *Create organizational play yards.* In order to break out of the mold of the current organizational thinking, it is often necessary to separate people or teams from the mainstream. If you want a team to develop creative solutions to a problem, provide them a work area removed from the day-to-day grind of production. In addition, create teams with diverse schools of thought. It has been shown that diverse opinions and devil's advocates within teams lead to more divergent thinking and subsequently more creative problem solutions [3].

2. *Hold people accountable.* Creativity without accountability is worthless. Even if teams are allowed to free-form problem solve, and even if they don't actually solve the problem they initially set out to solve, results of the problem-solving effort must be captured and reported. In this manner, teams are motivated to use resources wisely and to work harder to achieve results.

3. *Push the envelope.* Challenge people and teams to solve problems that seem unsolvable on the surface. It is amazing the amount of innovation that can result when teams are presented with stretch goals. Push beyond the idea that it's impossible to solve a problem to the idea that it's just a matter of time until it's solved.

4. *Encourage people to play multiple roles.* Innovative ideas must pass through many stages before they can come to fruition. Someone must initially be innovative and maybe even break a few rules.

Someone else must recognize the value of the idea and help obtain the resources necessary to push it along. Finally, someone must facilitate the disparate groups of people that have to come together to take an idea from the drawing board to the production floor. Encourage employees to look for opportunities to play any and all of these roles. In this manner everyone helps everyone else out, teamwork is enhanced, and innovative solutions become reality.

Conclusion

Whether a problem lends itself to a structured solution or a free-form solution, certain steps have to take place. The problem must be packaged in chunks small enough to tackle and then each of these chunks must be clearly defined. We have to know what problem we're trying to solve if we are to arrive at an optimal solution. We must also break out of the mold of thinking that tells us there is one and only one right solution to a problem. We should help our teams push beyond our innate mental blocks to creativity to develop innovative problem solutions. We have to foster this type of thinking environment by giving our employees the resources and the incentives to think creatively.

References

[1]   Whetton, D., and K, Cameron, *Developing Management Skills*, Reading, MA: Addison-Wesley, 1998.

[2]   Dauw, D., *Creativity and Innovation in Organizations*, Dubuque, IA: Kendall Hunt, 1976.

[3]   Nemeth, D., "Differential Contributions of Majority and Minority Influence," *Psychological Review*, 1986.

13

Time management

Introduction

Now we turn our attention to time management. It is well and good to employ the most up-to-date strategies to solve problems and accomplish project tasks, but all this is for naught if we, as project team members, are unable to perform these functions in a reasonable timeframe. Of the three traditional triple constraints in project management, time may be the one most difficult to manage. We have no way of altering the passage of time, we have no way of buying more time, and we have no way of regaining or recovering lost time. Once it's past, it's past. The best we can hope for is to manage our time wisely and efficiently. Project managers function as the hub of activity for a project, thus the project demands on their time are typically greater than anyone else involved. A problem arises here, because the project manager is expected to drive the time resources of the rest of the team. If he or she is unable to manage his or her own time resources, then he or she will not be able to successfully help team members manage theirs. Sophisticated project planning software exists which aids a project manager in scheduling project tasks, but rarely if ever are ad hoc phone calls, long lunches, or casual office banter figured into these plans. Not only that,

occurs, it should be structured so that the project manager gets the point of the conversation quickly, rather than having to wade through extraneous details before he or she gets to the point. Delegation is also a communication tool that can be used by the project manager to make better use of his or her time. The project manager should delegate tasks, meetings, and calls to team members who can handle them just as well and save his or her time for those tasks, meetings, and calls that only he or she can handle.

Stress

Now let's shift gears slightly and look at a topic closely related to time management, namely, stress management. Stress can be defined as a response to the stimuli from the environment which place demands on the body. Stress responses manifest themselves in different ways in different people, so it is difficult to determine when physical responses are due to stress and when they are due to other stimuli. One way to detect stress is to look for the three stages of reaction to stress by the body [3]. First, a person feels an adrenaline rush and either feels the need to flee from the stress or fight against it. Second, the body adapts to the stress stimuli and resists it. Finally, the body exhibits an exhaust reaction which is a return of the first state, but it is manifested through other vehicles such as headaches, ulcers, and heart attacks.

What stimuli in the workplace produce this stress reaction? The causes are as numerous as the number of stressed-out people in the work force. However, the following list is one that should encompass many common stimuli that most of us can identify with the following:

- Changes in job stability;

- Lack of recognition for accomplishments;

- Lack of clear direction in job tasks;

- A feeling of a lack of utilization of job skills;

- Too many demands on limited time available;

- Lack of control to do your job properly.

Once you understand your body's response to stress and those environmental triggers that cause stress, you should learn to control the response in yourself and help other team members control their responses if you recognize stress in them. There are several ways to try to alleviate or at least temper stressors in the environment and your response to them [4]:

- *Chill out.* The attitude you take toward work can influence what situations create stress for you. Find ways to make your daily tasks less of a drudgery and more of an enjoyable challenge. Try taking a positive outlook on the work ahead of you. A positive spin can help calm potentially stressful situations.

- *Manage your tasks.* Now we tie back to the time management discussion from above. One of the greatest stress relievers is to ward it off in the first place before it ever starts. Organize your work life, concentrate on the issues that are truly important and that you can truly affect, and utilize your time wisely. Manage your time so that deadlines don't truly become drop-deadlines.

- *Schedule leisure time.* That may sound like an unusual juxtaposition of terms, schedule and leisure, but it is important to delegate time for leisure and rest. Our bodies and minds are not designed to work constantly. Leisure time is often the first time to get squeezed when we do a poor job of time management. If we continue to push leisure time to the side, we will ultimately pay the price in the physical manifestation of workplace stress.

Conclusion

Time is a limited commodity and it is unusual in the fact that once we use it, it is gone. We get 24 hours per day, no more and no less. Given the nature of time we must learn to manage it wisely if we are to be efficient in performing our work tasks. In order to make every minute that we have available count, we must employ some sort of project planing system that will allow us to group tasks by importance and to prioritize those tasks. A project manager must use his or her authority to delegate effectively so that everyone on the team, project manager included, works on the tasks most suited to his or her abilities and expertise. The project manager must also track team progress on tasks so that if a task is falling behind schedule, measures can be taken immediately to bring it back in line, thus saving time that might later be wasted if the problem is allowed to persist and grow.

Closely related to time management is stress management. There are few things more stressful in the workplace than the feeling that you have too many tasks to accomplish in a given period of time. The obvious solution to stress relief is to remove yourself from the stimuli in the environment that are producing stress responses. Time management is one of the best tools to employ to relieve this stress as it allows you to slot tasks into time slices that are of the appropriate length. Time management can also be utilized to reduce stress by planning leisure time into your schedule so that you have the opportunity to refresh yourself both mentally and physically. Inattention to these opportunities to recharge may lead to a variety of health problems and potentially death.

References

[1] Pennypacker, J. (ed.), *Principles of Project Management,* Sylva, NC: PMI Publications Division, 1997.

[2] Douglas, M., and D. Douglass, *Manage Your Time, Manage Your Work, Manage Youself,* New York: Amacom, 1980.

[3] Greenwood III, J., and J. Greenwood Jr., *Managing Executive Stress: Systems Approach,* New York: John Wiley and Sons, Inc., 1979.

[4] Giammateo, M., and D. Giammateo, *Executive Well-Being—Stress and Administrators,* Reston, VA: National Association of Secondary School Principals, 1980.

14

Auditing

Introduction

Our last miscellaneous topic deals with project auditing. In addition to all the other project elements a project manager must keep up with, project audits are not the least of his or her worries. Actually, it behooves the project manager to institute his or her project data gathering and control system early in the project life cycle so that the appropriate data might be captured as it is generated, thus reducing the stress of ad hoc data gathering when the auditors show up. Project managers should plan for auditing all along the way of the project's life and be prepared to account for any and all elements on the project WBS.

Auditing defined

Basic Auditing Concepts of the American Accounting Association defines auditing as "a systematic process of objectively obtaining and evaluating evidence regarding assertions about economic actions and events to ascertain the degree of correspondence between those assertions and established criteria and communicating the results to interested users" [1].

In general, audits are conducted to determine the following:

* Whether activities and programs being implemented have been properly authorized;

* Whether activities and programs are being conducted in a manner contemplated to accomplish the objectives intended by an organization's governing body;

* Whether activities or programs are performed in an efficient and effective manner;

* Whether tasks and functions are performed in compliance with applicable laws, policies, and procedures;

* Whether revenues are being properly collected, deposited, and accounted for;

* Whether resources, including funds, property, and personnel are adequately safeguarded, controlled, and used in a faithful, effective, and efficient manner;

* Whether during the course of audit work, there are indications of fraud, abuse, or illegal acts;

* Whether there are adequate operating and administrative procedures and practices, systems, or accounting internal control systems and administrative controls which have been established by management.

Types of audits

The most common application of auditing is a financial report audit. The objective of a financial report audit is to enable the auditor to express an opinion as to whether the financial report is prepared, in all material respects, in accordance with an identified reporting framework (Auditing Standard ISA 200, *Objective and General Principles Governing an Audit of a Financial Report*). Other types of audit include compliance audits, performance audits, comprehensive audits, environmental audits, and internal audits. These different types of audit encompass the following:

- *Financial report audit.* Examines an organization's financial records (the most common type of audit).

- *Compliance audit.* Involves obtaining and evaluating evidence to determine whether certain financial or operating activities of an entity conform to specified conditions, rules or regulations (most relevant to the public sector).

- *Performance audit.* Involves obtaining and evaluating evidence about the economy and effectiveness of an entity's operating activities in relation to specified objectives (closely related to internal auditing).

- *Comprehensive audit.* Usually occurs when an auditor undertakes a range of audit and audit-related services (more common in the public sector and includes financial, compliance, and performance audits).

- *Environmental audit.* Covers environmental matters which may have an impact on the financial statements (the activities of a business may impact on reported assets and profits due to their effects on the environment and environmental regulations).

- *Internal audit.* Refers to any of the activities listed above, but conducted by auditors employed by the organization (internal audit is a management tool used by the organization to enhance internal control).

The auditor of information that is the responsibility of one party (such as management), but is for use by other parties (such as shareholders, customers, suppliers, governments), requires technical competencies in three main disciplines:

1. Accounting (such as financial or environmental accounting);

2. Information systems;

3. Cognitive decision making.

The activity-based risk evaluation model of auditing (ABREMA) integrates the three descriptive concepts of the audit objective, information misstatements, and audit stages, with the two theoretical concepts of cognitive decision making and audit risk.

Concept of cognitive decision making

It is assumed that auditors, when performing the critical decisions referred to in each of the audit stages, undertake a structured set of activities that correspond with the concepts of human information processing (HIP) theory as they relate to cognitive decision making. The audit activities (and their corresponding HIP activities) include the following:

1. Planning (hypothesis generation);

2. Evidence gathering (information search);

3. Evidence evaluation (information evaluation);

4. Decision making (choice).

Concept of the audit objective

It is assumed that the objective of the audit of any information prepared by a particular party, including financial statement information prepared by management, is to gather and evaluate evidence of sufficient quantity and appropriate quality to form, and to communicate to the users of the information, an opinion on the underlying assertions made by the party preparing the information, for the purpose of adding credibility to those assertions.

Concept of information misstatements

All misstatements of information (including misstatements of financial statement information) can be categorized into one of three categories:

1. *Completeness.* If an item which should be included is not included, then the information is not complete.

2. *Validity.* If the information includes an item that should not be included, the information is not valid.

3. *Accuracy.* If the information includes an item that should be included, but it is not included accurately, the information is not accurate.

Concept of audit stages

There are five sequential audit stages in an audit engagement that correspond to the five critical decisions made during the course of the audit. The stages (and corresponding critical decisions) are as follows:

1. Client acceptance and retention stage (the decision as to whether to accept or reject a prospective client or to retain or relinquish an existing client;

2. Audit planning stage (the decision as to the appropriate audit approach);

3. Control testing stage (the decision as to the extent to which reliance on information system controls continues to be appropriate);

4. Substantive testing stage (the decision as to the extent of misstatements in the data underlying the information presented to information users);

5. Opinion formulation stage (the decision as to the appropriate audit opinion on the information presented to information users). Concept of audit risk.

The evaluation of audit risk (the risk of a material misstatement in the audited information), or its components, is required during each of the five audit stages to assist in resolving the critical decisions relevant to each stage.

Although some of the above concepts are neither new, nor particularly controversial, the manner in which these concepts are brought together into a single, integrated model provides the basis for a structured, modular approach to the teaching of auditing as well as being a novel contribution to the conceptual framework discussion [2] (Table 14.1).

TABLE 14.1   A Structured Approch to Teaching Auditing

	AUDIT STAGES				
AUDITOR ACTIVITIES	Client acceptance/ retention	Audit planning	Control testing	Substantive testing	Opinion formulation
PLANNING	Strategic planning	Tactical planning	Operational planning	Operational planning	Operational planning
EVIDENCE GATHERING	Preliminary knowledge of business	Detailed knowledge of business	Effectiveness of controls	Substantive-ness of underlying data	Consistency of final information with auditor's knowledge of business
EVIDENCE EVALUATION	AR*1 ~AR1	DR2 = AR/IR2 x CR2	CR2 ~ CR3	AR*4 ~ AR4	AR*5 ~ AR5
DECISION MAKING	Accept/retain (or reject/ relinquish)	Audit approach	Continued reliance on controls	Conclusion on individual data items	Audit opinion

Steps of an audit

Auditors must engage in the following steps before beginning an audit [3].

Step 1: Evaluate management integrity

The first thing the auditor must do is to evaluate the integrity of management. The auditor should be satisfied that the management of the firm being audited can be trusted. If the auditor discovers that the management lacks integrity or is disreputable, they are faced with a number of problems and potential problems. For example, it may be that the reason the auditor has been approached to complete the audit is that previous auditors have discovered material misstatements in the accounts and indicated they would qualify them. Management may have sacked them and turned to another firm in the hope that they would not discover the misstatements and issue a clean report. Alternately, if the auditor finds that management can't be trusted totally, but decides ultimately to accept the audit, the extent of their testing procedures may be increased to ensure that they discover any material errors or irregularities in the financial statements. The auditor can assess the integrity of the management of the firm in two main ways. First, by contacting the previous auditors of the firm (if the firm is not being audited for the first time) and second, by making inquiries of other third parties. These third parties could be such people as bankers who have dealt with the firm, creditors, lawyers, and others.

Step 2: Identify special circumstances and unusual risks

The second step in this preplanning process that leads to the preparation of the audit engagement letter is the identification of any special circumstances or unusual risks that may affect the nature and scope of the audit. For example, auditors need to be aware of the people who are going to read and rely on the financial statements of the firm being audited. If these users have special needs, or are special in themselves, then the auditor may be obliged to complete the audit and write the audit report in a special way to insure the users can understand the contents of the financial statements and what the auditor is saying about them. Equally, the auditor needs to identify the level of risk involved in undertaking the audit. For example, the firm requesting the audit may be in financial distress which could lead to its collapse. In today's world where many people and organizations use the courts to seek remedies, the auditor may be exposed to costly litigation if he or she takes on a risky firm that collapses. Stakeholders in the firm may sue the auditor for damages. It may be wiser for the auditor to eliminate this risk by rejecting any audits where the potential for this type of risk exists. Also associated with this step is the evaluation of the auditability of the firm. The financial records and internal controls may be in such a poor state that the auditor is aware from the outset that he or she will not be able to obtain the evidence he or she needs to complete her audit report. In this case, it would be best for the auditor to decline the audit.

Step 3: Evaluate independence

The third, perhaps most important, step in the process of deciding whether to issue an engagement letter is to determine if the acceptance of the audit may impair the auditor's independence. In order to ensure that the interests of the client are at all times protected, the auditor must be perceived as being independent as well as actually being independent.

Step 4: Assess competence to perform audit

The fourth step involves assessing the firm's competence to performing the audit. Thus, before accepting an engagement, auditors should determine whether they can complete the engagement with due care and professional competence. Generally, this involves identifying key members of the audit team and considering the need to seek assistance from experts.

Step 5: Determine ability to use due care

Even if the auditors have decided at this point that they are able to act independently for the potential audit client, it is important that they determine whether they have the ability to conduct the audit with due care. That is, do they have the expertise? Have they access to consultants and specialists, if needed? Are they able to complete the audit in a timely fashion? Will they be able to complete the audit within the budgeted audit figure expected by the client? All of these questions need to be addressed to ensure that, if undertaken, the audit will be completed with due care.

Step 6: Prepare engagement letter

The final step prior to preparing the engagement letter is to ensure that the appointment, if made, is done in accordance with the various professional, ethical, and statutory requirements. Once this step is complete and the auditor is satisfied that he or she can accept the appointment, an engagement letter can be prepared and forwarded to the client. It is important that the nature and scope of the audit is clearly stated in the letter. This ensures that both the client and the auditor have no misunderstandings on what the auditor must do to properly discharge his or her duties.

Once the audit engagement letter is prepared and sent, the auditor must then develop an overall plan of the audit. An audit plan is critical to the conduct of an audit. It forces the auditor to think through and detail each of the following steps needed leading to the preparation of the audit:

1.   Understanding the client's business and industry;

2.   Performing analytical review procedures;

3.   Assessing materiality levels;

4.   Assessing audit risk;

5. Developing preliminary audit strategies;

6. Understanding the entity's internal control structure.

No auditor can satisfactorily complete the audit of an entity without a comprehensive understanding of the entity's business activities (i.e., what and how it trades) and the industry within which the entity operates.

Closely associated with gaining an understanding of the client's business and industry is the performance of analytical procedures. In auditing, these procedures are used to achieve three main objectives. First, to help the auditor plan the nature, timing, and extent of the audit. Second, as a part of the substantive testing procedures which, for example, are used to obtain evidence about the assertions made in the financial statements. And third, at the end of the audit, to gain the broad-picture view of the client's financial statements.

The completion of this analytical review phase of the audit plan includes six steps:

1. Identifying the computations, comparisons, or relationships to be investigated;

2. Estimating probable outcomes;

3. Performing comparisons;

4. Analyzing data and identifying significant differences;

5. Investigating significant unexpected differences;

6. Determining effects on audit planning.

After obtaining an understanding of the business and completing the analytical procedures, the auditor must next make a judgment about the materiality levels and consider the associated problem of audit risk. Once preliminary judgments are made about materiality levels, and once the audit risk is assessed, the auditor is then in a position to determine what is sufficient and appropriate audit evidence upon which he or she can base her audit opinion.

The international auditing standard ISA 320 defines materiality as "…in relation to information, that information which if omitted, misstated or not disclosed separately has the potential to adversely affect decisions about the allocation of scarce resources made by users of the financial report or the discharge of accountability by the management, including the governing body of the entity."

This is a rather wordy definition but what it essentially means is that if there is an error (deliberate or otherwise) in the financial information of an entity, and that by being there it influences the decisions shareholders may make about their investments in the entity, then that error is regarded as material. Given this, the auditor needs to assess, quantitatively and qualitatively, what may lead to material misstatements in his or her planning phase so that he or she can establish what evidential matter he or she needs to obtain to satisfactorily complete his or her audit. There are two levels of materiality, (a) materiality at the financial statement level, and, (b) materiality at the account balance level. Any misstatements that prevent the financial statements from being presented fairly is a material misstatement. When discussing materiality at the account balance level, the term *materiality* refers to the amount of misstatement that could affect a user's decision. Generally, there is a direct relationship between the size of an account balance and the amount of evidence needed for assertions related to the account. On the other hand, there is an inverse relationship between materiality and the amount of evidence needed. That is, the smaller the materiality, the more evidence that is needed to obtain reasonable assurance that misstatements do not exceed that level. If accounts had to be perfectly accurate, auditors would have to check every transaction (even then it is unlikely the accounts would be perfectly accurate). Materiality, however, allows the auditor to check only a sample of items, as they will be relying on statistics and the fact that small items are unlikely to contain significant errors. In this way the number of items checked in an audit is a small percentage of the total transactions, and the audit can be completed at an acceptable cost.

Currently, most auditors use risk analysis to help them plan their audit and establish the nature and level of audit evidence they need to collect so they can issue their audit report. The aim of this risk analysis is to determine the audit risk for the client under review. The international standards dealing with materiality and audit risk (ISA 320 and 400) describes audit risk as the risk that the auditors may express inappropriate opinions on financial information that is materially misstated. In practical terms what this means is that the auditor attempts to quantify the risk of declaring the financial statements to be clean, when in fact they are not. So, if the auditors wish to be 95% certain that the opinion they express is correct, then they need to insure that their audit procedures will result in a 5% audit risk.

Audit risk is calculated through assessing the level of inherent risk, control risk, and detection risk. The only way the auditor can control the level of audit risk is through detection risk. That is, the level detection risk can be changed by varying the nature, timing, and extent of substantive tests performed on any item included in the financial statements. As a general rule, the lower the level of desired audit risk, the higher the level of detection risk. The relationships between these three components of audit risk can be stated mathematically as $AR = IR \times CR \times DR$, where AR is audit risk, IR is inherent risk, CR is control risk, and DR is detection risk. Knowing that the relationships can be stated this way means that the auditor can determine any one of the types of risk if a value is known for the remaining three. This knowledge allows the auditor better plan his or her audits.

Armed with an understanding of the various risks associated with the audit, and knowing the critical nature of control risk in assessing audit risk, the auditor can develop a preliminary audit strategy.

Internal auditing

Internal auditing is an independent appraisal function established within an organization to examine and evaluate its activities. Thus, the internal auditor is employed within an organization but is independent of management. The internal auditor has responsibilities to evaluate all aspects of the organization's activities [4].

The internal audit function includes the following:

- Reviewing and appraising the soundness, adequacy and application of accounting, financial and operating controls;

- Ascertaining the extent of compliance with established policies, plans and procedures;

- Ascertaining the extent to which the company assets are accounted for and safeguarded from losses of all kinds;

- Ascertaining the reliability of accounting and other data developed within the organization;

- Appraising the quality of performance in carrying out assigned responsibilities.

Internal audit operates to do the following:

- Insure adequate internal control exists;

- Review the reliability of the records;

- Prevent and detect fraud;

- Monitor reporting procedures;

- Policies, procedures, and legal requirements are followed;

- Assets are safeguarded;

- Resources are used economically;

- Objectives are achieved in a timely manner.

The Institute of Internal Auditors, an international body, has issued five general standards covering the work of internal auditors:

1. Independence;

2. Professional proficiency;

3. Scope of work;

4. Performance of audit work;

5. Management of internal audit department.

Some key checklist areas that should be of concern to internal auditors include the following:

- *Independence.* Internal auditors should be independent of the activities they audit.

- *Compliance with standards of conduct.* Internal auditors should comply with professional standards of conduct.

- *Knowledge, skills, and disciplines.* Internal auditors should possess the knowledge, skills, and disciplines essential to the performance of internal audits.

- *Human relations and communications.* Internal auditors should be skilled in dealing with people and in communicating effectively.

- *Due professional care.* Internal auditors should exercise due professional care in performing internal audits.

- *Scope of work.* The scope of internal auditing should encompass the examination of the adequacy and effectiveness of the organization's system of internal control and the quality of performance in carrying out assigned responsibilities.

- *Reliability and integrity of information.* Internal auditors should review the reliability and integrity of financial and operating information and the means used to identify, measure, classify, and report such information.

- *Compliance with policies, plans, procedures, laws, and regulations.* Internal auditors should review the systems established to ensure compliance with those policies, plans, procedures, laws, and regulations which could have a significant impact on operations and reports and should determine whether the organization is in compliance.

- *Safeguarding of assets.* Internal auditors should review the means of safeguarding assets and, as appropriate, verify the existence of such assets.

- *Economical and efficient use of resources.* Internal auditors should appraise the economy and efficiency with which resources are employed.

- *Accomplishing established objectives and goals for operations or programs.* Internal auditors should review operations or programs to ascertain whether results are consistent with established objectives and goals and whether the operations or programs are being carried out as planned.

- *Performance of audit work.* Audit work should include planning the audit, examining and evaluating information, communicating results, and follow up.

- *Planning the audit.* Internal auditors should plan each audit.

- *Examining and evaluating information.* Internal auditors should collect, analyze, interpret, and document information to support audit results.

◆ *Communicating results.* Internal auditors should report the results of their audit work.

◆ *Following up.* Internal auditors should follow up to ascertain that appropriate action is taken on reported audit findings.

Table 14.2 notes the differences between internal and external auditors.

External auditors are independent of the company they are auditing. They are engaged to express an opinion as to the truth and fairness of the financial report. They report to members of the company.

Internal auditors are employees of the entity they audit. They are involved in an independent appraisal activity called internal auditing within an organization as a service to the organization. The objective of internal auditing is to help the management of an entity in the discharge of its responsibilities.

Other key differences between internal and external auditing include the following:

◆ Internal auditing is mainly to assist management whereas external auditing is mainly for the entities' members and shareholders.

◆ The role of internal audit is generally much wider than external audit. It may involve complex systems review, internal checking, forensic investigations, internal appraisals of operations, and financial planning or even a financial report audit.

The two tasks, however, are complementary. The internal audit function is invaluable because of the assistance it can provide management. The external audit is also important, because performance of an audit by an

TABLE 14.2   Internal and External Auditing Differences

INTERNAL AUDITORS	EXTERNAL AUDITORS
Audit continuously	Audit at year end
Report to management	Report to external sources
Audit all management systems	Audit financial systems
Are concerned with effective, efficient, economic achievement of strategic goals	Are concerned with accurate and complete reporting of financial status

external party provides users with assurance that financial statements are fairly stated.

Auditing terms

At this point it may be a good idea to review some working definitions of terms that are common within the field of auditing.

Independence: An impartial attitude free of any interest which might be regarded, whatever its actual effect, as being incompatible with integrity and objectivity. The auditor must be without bias with respect to the client, since otherwise he or she would lack the impartiality necessary for the dependability of his or her findings. Independence requires intellectual honesty and judicial impartiality that recognizes an obligation for fairness not only to management and owners of a business, but also to creditors and those who may otherwise rely on the independent auditor's report. The essence of this statement is that it is not enough that the auditor be independent, but that he or she also be seen to be independent, often referred to as actual independence and perceived independence.

Materiality: A quality of information which if omitted, misstated or not disclosed separately has the potential to adversely affect decisions about the allocation of scarce resources made by users of the financial report or the discharge of accountability by the management, including the governing body of the entity. ISA No. 320 (*Audit Materiality*) of July, 1994, states: "(3) Information is material if its omission or misstatement could influence the economic decisions of users taken on the basis of the financial statements.... (12) In evaluating the fair presentation of the financial statements, the auditor should assess whether the aggregate of uncorrected misstatements that have been identified during the audit is material.... (15) If management refuses to adjust the financial statements and the result of extended audit procedures do not enable the auditor to conclude that the aggregate of uncorrected misstatements is not material, the auditor should consider the appropriate modification to the auditor's report...." The concept of materiality involves the paradox that the auditor made his or her materiality decisions, on behalf of the users of the financial statement. This is a rather wordy definition but what it essentially means is that if there is an error (deliberate or otherwise) in the financial information of an entity, and that by being there it influences the decisions shareholders may make about their investments in the entity, then that error is regarded as material. SEC Staff Accountant Bulletin (SAB) No. 99 provides guidance for preparers and independent auditors on evaluating the materiality of misstatements in the financial reporting and auditing processes by summarizing and putting in perspective certain GAAP and federal securities laws as they relate to materiality.

Consistency: The adherence to standards, format, vocabulary, terms, appearance, objectives and goals.

Comparability: An attribute of a comparison method which properly reflects similarities and differences between a subject and its potential comparables. Information about a particular company is more useful if an investor can compare it with similar information about other companies and with similar information about the same company for some other time period. We used to call this comparing apples to apples (that is, before Apple became a computer!) The purpose of comparison is to detect and explain both similarities and differences. High-quality accounting requires accounting for similar transactions similarly and accounting for different transactions and circumstances differently.

Representational Faithfulness: Financial information should be presented in a way that is not misleading. High-quality financial statements should reflect the real economics of the transaction being reported. In the United States, we refer to this concept as representational faithfulness. A map's representational faithfulness may be determined by how well the map describes the coastline. In the same way, a financial statement's representational faithfulness may be evaluated by how well it represents the economic resources and obligations of the company, and by how well the transactions and events that change those resources and obligations are described. For example, suppose that a company reflects future costs in its income statement by setting up reserves in a profitable year and then reducing those reserves in a bad year when the costs are actually incurred. Investors are unable to see the real economic results of the business. Accounting standards that permit this practice lack high quality because they don't exhibit representational faithfulness.

The bottom line in representational faithfulness is that transactions need to be accounted for based on their true economics rather than just on their form. And we need companies, their auditors, financial analysts, and investors to focus on this as an integral objective of high-quality, transparent reporting, rather than trying to add bells and whistles to a transaction purely to achieve a different accounting treatment. If you try to paint stripes on a horse just so you can call it a zebra, watch out when the rain falls! And you better make sure that horse doesn't come back and kick you for trying to portray it as something that it's not.

Efficiency: The ability to perform some task given certain resource constraints. Audits can measure the efficiency of utilization of human, financial, and other resources, including examination of information systems, performance measures and monitoring arrangements, and procedures followed by audited entities for remedying identified deficiencies.

Effectiveness: The ability to perform some task and achieve previously stated goals based on previously defined metrics. The effectiveness of an auditing program can be measured by how well the program supports the objectives and goals of the organization. An effective auditing process must include follow-up inspections. The effectiveness of an auditing program is dictated by the following:

- Adequacy of the auditing program;

- Proper planning;

- Supervision and review of the audit work performed.

Audit tests: Comparisons of process results against documented procedures and expected results of those procedures.

Audit evidence: Audit evidence is the information auditors use to meet audit objectives and indirectly to test whether operating objectives are being met. Audit evidence must be sufficient (enough) and competent (quality). Other attributes that audit evidence should possess include freedom from bias, objectivity, and relevance.

Examples include the following:

- Physical evidence;

- Testimonial;
- Documentary;
- Analytical.

The rules and standards of audit evidence dictate that the evidence gathered during the audit be the following:

- *Sufficient, persuasive.* This requires the exercise of good professional judgment. Audit evidence from the most to least persuasive include: physical examination, externally prepared, observations, inquiries and testimonial.
- *Competent.* Obtaining the best quality of evidence available.
- *Useful.* Evidence supporting goals and objectives.
- *Relevant.* Evidence needs to be logical and sensible relative to the audit finding.

Various techniques are used to gather audit evidence:

- Physical examination;
- Confirmation;
- Observation;
- Recalculating;
- Reconciliation;
- Inquiry;
- Inspecting;
- Analytical procedures;
- Detail testing.

How does an auditor know how much or what kind of evidence to obtain? Experience, professional judgment, and common sense all play a key role in making this determination.

Auditors' Report: A written audit report is issued for each engagement. This report includes an opinion on the effectiveness of the system of internal control, significant audit findings, and agreed-upon corrective procedures. At a closing conference, the auditee has an opportunity to suggest changes to the wording contained in the report. The auditee may also provide written comments on the significant audit findings. These suggested changes and comments are integrated into the final report. The auditor should supply sufficient information for the customer so that a solution to identified problems may be secured. This is particularly important when alternative solutions are available, or when additional assistance is needed. Also, implications and recommendations for those involved should be included.

The first thing a project manager can do to ensure a successful audit is to understand what a project audit is and what it will involve. Project audits are not necessarily limited to financial matters and they may or may not be performed by financial personnel. The purpose of an audit is, depending on when it is conducted, to help a project achieve its stated goals or to determine if the project met its stated goals. Audits will cover the general project areas of cost, schedule, quality, and risk. The report will, again depending on when the audit is conducted, examine where the project is in relation to where it should be, what risks are ahead, suggestions to handle these risks, a report on how past risks were handled, and any other recommendations the auditor may have. The project audit will consist of the following steps:

1. *Initiation:* Scope, purpose, and methodology are determined;

2. *Baseline:* Establishment of standards against which to measure the project;

3. *Data collection:* Self-explanatory;

4. *Analysis:* Overview of findings based on data collection;

5. *Reporting:* Full-blown audit report with findings and recommendations;

6. *Termination:* Development of action plans and documentation of lessons learned.

Project managers should also understand that their project planning should include planning for periodic project audits. This means that the following information should be collected and organized all during the life cycle of the project:

- Expected and actual costs of WBS tasks;
- Expected and actual time consumption of WBS tasks;
- Human resource utilization;
- Change control records;
- Original scope, mission, and vision statements of the project;
- Documentation of management support for the project;
- Risk analysis and mitigation strategies for each risk;
- Documentation of project achievements.

Conclusion

In general, the project manager needs to insure that project record keeping is timely and accurate. If things are going badly, the documentation should reflect that. If things are going well, the documentation should reflect that also. Including audit preparation in the initial planning phases of the project should address the issue of data availability when a project audit occurs.

I believe audit planning in total supports a project manager's control objectives. If audit data is collected accurately and in a timely fashion, not only will it ease the audit process, but it will enable the project manager to have a clear, up-to-date picture of project performance. It will allow him or her to spot problems as soon as they arise and deal with them expeditiously. Project control boils down to having current project data available and in a useful format and this goes hand in hand with the requirements of a project audit.

References

[1]   http://www.theiia.org/ecm/guide-pc.cfm?doc_id=696.html.

[2]   http://www.abrema.net/abrema.

[3]   http://www.ican.org.np/public_html/auditing/lecture_1.htm.

[4]   Marshall, D., and W. McManus, *Accounting: What the Numbers Mean,* Boston, MA: McGraw-Hill, 1996.

15

Future directions for project management

Interest in project management has grown phenomenally in the last 10 years. Organizations are increasingly making use of project management methodologies and membership in organizations such as PMI is growing exponentially. The field of project management itself is evolving, from a purely technical, methodical approach to projects, to a more personal, client-oriented approach. Just reference PMI's *Project Management Body of Knowledge* (PMBOK) and the changes in that document over the last few years and you will see the increased importance placed on areas such as communication and project human resources. In this book I have also tried to emphasize the importance of the human side of project management such as personality awareness, motivational techniques, and communication both inside and outside the project team. The remainder of this chapter will examine specific evolutionary trends of project management and what project managers can do now to prepare themselves for these developments.

Project office growth

The growing use of project methods will drive organizations to establish more and larger project offices that will provide staffing and methodology to help projects be more successful. The project office also allows for a single point of contact in a matrix organization with many simultaneous, complex projects. Another development will be the increased use of program offices and program managers in which many related projects are aggregated underneath a common manager. The development of program offices brings organizations one step closer to realizing a fully projectized organization where the tasks of the organizations are aligned by process rather than function.

Virtual teams

The use of virtual teams is widespread now and will become even more widespread in the near future. The globalization of the world's economies makes this a natural occurrence. The shortage of technology workers is also driving the push for virtual teams and telecommuting. It is impractical for a company to try to draw workers from across the globe and get them all to settle in one spot. Rather, it makes more sense to utilize the advances in telecommunications and connect a group of workers virtually. This also allows organizations to extend the hours over which they have employees working. If a company has employees across the globe, they can have productive work going on 24 hours a day.

Collaboration tools

Technological leaps forward in both telecommunications and software development have resulted in a number of collaboration tools that allow virtual teams to share information easily. Very few places exist on the Earth anymore that cannot be reached via some means of electronic communication. Video conferencing tools now reside on desktops rather than in expensive, specially designed conference rooms. Groupware is available that allows team members to share documents and contribute to work jointly, rather than in sequence. We are beginning to see the merger of voice and data communication devices, such as Web services over digital wireless phones. These innovations and advancements will continue for the foreseeable future and will continue to make physical position of team members less and less important.

Joint application development

Competition, rapid technological advances, and shifts in the global economy have forced organizations to become more nimble. Reaction time to market changes must be shortened and time to market with new products and services must be shortened. All this must happen without impacting quality to the customer. One way to shave time off development is with joint application development. This involves bringing all the players involved with a situation together at once to collect and document requirements. In this way, requirements can be gathered more quickly than by doing individual interviews, all participants in the project can be made aware of the requirements simultaneously, and a sense of team is developed among participants from disparate elements of the organization.

Outsourcing

The future will see people become less devoted to a particular organization and more devoted to their skill or craft. As this trend continues, organizations will find that it is more beneficial for them to outsource certain functions than to try to retain enough qualified personnel to perform the functions in-house. In addition, outsourcing takes the burden off the organization of trying to stay current with the latest technology. Rather than pouring resources into training personnel and upgrading equipment, organizations can just go out on the open market and contract for the services they cannot provide for themselves. They can then obtain as high a level of expertise with a particular technology as they are willing to pay for.

Risk management

Risk management has always been an important element of project management, but it will become even more important as the profession evolves in the future. Shortened cycle times for projects and increasing complexity of projects increases project risk. Project process will be scrutinized more closely to eliminate risk. Vendors and clients will partner more closely to share risk more equally. Finally, organizations will place more emphasis on quality certification and process maturity certification, both for themselves and for their suppliers.

Project-management education

Education and certification of project managers will grow and will become more important. The number of PMP's certified by PMI has been steadily growing and more and more institutions of higher learning are offering degree programs in project management. As organizations utilize project management more fully, education and certification in project management will be in even greater demand than it is today.

Flexibility

The amount of work to be done in the future will not necessarily be less than it is right now, but it will certainly be different. We are in the midst of a revolution in the type of work people do and in the way they do their work. We saw this type of work revolution in the nineteenth century as we moved from a mostly agrarian society to a mostly industrial society. Today we are seeing a shift in work from industrial production to service delivery. The need for flexibility is one driving factor pushing organizations to projectize their work more fully. Project managers need to be flexible enough to work on one project this month, then move to another next month. Organizations who will be successful in the future will be ones who can organize, execute, disband, and reorganize projects quickly.

The value of people

It may seem odd that we talk about organizing and dismantling projects and then talk about the value of people in the same breath. Doesn't the projectized environment just treat people as cogs of the organization to be assembled, dismantled, and then reassembled in new ways? Actually, it is this sort of thinking that leads to trouble when organizations employ projects and project management. They must realize that the project structure is not what breeds success, rather it is the fact that people's skills are being utilized more effectively by allowing them to work in a project structure. The project structure places value on the skills a particular individual can bring to the project, thereby giving that individual incentive to continually seek to improve his or her skills. The project structure is worthless without skilled personnel with which to staff it.

Conclusion

We ended the first chapter of this book by stating some truisms about succeeding as a project manager and also succeeding as an organization employing project management methodology. I think it is useful to look back at those points here in light of what I have outlined for the future of project management.

1. *Do your homework.* Keep yourself apprised of the most current best practices in project management. Don't just fly by the seat of your pants. You're certain to fail if you take this approach. This will be even more true as project management moves forward. Reduced project cycle time leaves less room to learn as you go. Project managers and organizations in general will need to keep themselves abreast of what the competition is doing and try to improve on best practices in the industry.

2. *Sweat the details or bleed the implementation.* It is human nature to want to immediately rush to the end goal of a journey. As a project manager you must fight the urge to ignore the upfront details of planning just to get to the execution stage of the project. Do the right things right the first time and you will reap the dividends by the end of your project. Once again we see the essence of speed with accuracy. As organizations employ project management practices more fully they must make a concerted effort to capture the lessons learned from one project and apply them to other projects. As organizations build this repository of knowledge, their

Appendix A:
Risk and opportunity assessment

The following instrument is a model that may be used to assess the risk and the opportunity presented by a particular project. Ideally the model should be utilized immediately after requirements definition is complete. The process of utilizing the model involves the following three steps:

1. *Risk evaluation.* A risk score is obtained by multiplying the risk probability by the predetermined risk impact for each risk question. The scores for all the risk questions are then added together to arrive at a total risk score. (Table A.1)

2. *Opportunity evaluation.* An opportunity score is obtained by multiplying the possible opportunity by the predetermined opportunity weight for each opportunity question. The scores for all the opportunity questions are then added together to arrive at a total opportunity score. (Table A.2)

3. *Total risk.* The total risk score is placed on the horizontal axis of the model's matrix and the total opportunity score is placed on the

vertical axis. The intersection of the two scores is plotted and the location of this point on the matrix determines the quality of the opportunity and the degree of risk of this particular project. (Table A.3)

TABLE A.1   Risk Assessment

QUESTION	IMPACT	PROBABILITY	TOTAL (I × P)
What is the level of customer commitment?	5		
1. Personnel and budget are assigned.			
2. Budget is assigned.			
3. Personnel are assigned.			
4. Neither personnel nor budget are assigned.			
How was the project timeline built?	4		
1. Start and end times are flexible.			
2. Start and end times were mutually agreed upon.			
3. Start and end times were set by the client.			
4. Start and end times were set by the client and cannot change without penalty.			
What is the expected project duration?	3		
1. Fewer than 3 months			
2. 3 to 6 months			
3. 6 to 12 months			
4. More than 1 year			
Do we have experience with this type of project?	4		
1. Project is exactly like one we have done before.			
2. More than half the project requirements are like a previous project.			
3. Between 10% and 50% of the project requirements are like a previous project.			
4. We have never done a project like this.			
What was our level of involvement in requirements definition?	3		
1. We developed the requirements.			

TABLE A.1 (continued)

Question	Impact	Probability	Total (I x P)
2. We assisted the client in requirements definition.			
3. We had the opportunity to comment on requirements already developed.			
4. We had no involvement in requirements definition.			
How many vendors and/or subcontractors will be involved in the project?	3		
1. None			
2. 1 to 2			
3. 3 to 5			
4. 6 or more			
What is the timeframe for responding to the initial customer requirements?	3		
1. No requirement			
2. Conservative timeframe			
3. Moderate timeframe			
4. Aggressive timeframe			
What is the maturity of the materials required to satisfy the project requirements?	2		
1. All required materials are mature.			
2. More than 70% of the materials are mature.			
3. 30% to 70% of the materials are mature.			
4. Less than 30% of the materials are mature.			
How wide is the project's geographic distribution?	2		
1. Confined to one region within the country			
2. Confined to one country			
3. Confined to two continents			
4. Global distribution			
What is the project manager's overall risk assessment of the project?			
1. Extreme			

TABLE A.1 (continued)

Question	Impact	Probability	Total (I × P)
2. High			
3. Moderate			
4. Low			

TABLE A.2   Opportunity Assessment

Question	Impact	Probability	Total (I × P)
How many corporate strategies are matched by this project?	5		
1. Less than 50%			
2. 50% to 66%			
3. 67% to 75%			
4. More than 75%			
What is the relative monetary value of the project?	4		
1. Low			
2. Moderate			
3. High			
4. Extreme			
How do the gross margins for the project compare to corporate target margins?	4		
1. Break even or less			
2. Up to 50% of target			
3. 50% to 100% of target			
4. More than 100% of target			
What impact will this project have on future business with this customer?	3		
1. No impact			
2. Future business possible			
3. Future business likely			
4. Future business assured			

TABLE A.2 (continued)

Question	Impact	Probability	Total (I × P)
How valuable is the experience and skills to be gained from project experience?	3		
1. Little improvement			
2. Significant improvement in existing skills			
3. Some new skill development			
4. Significant new skill development			
What is the impact on resource utilization?	3		
1. Significant drain			
2. Some resource drain			
3. Normal impact			
4. No drain			
How does the customer view in against our competition?	2		
1. The customer favors the competition and is negative toward us.			
2. The customer favors the competition and is neutral to us.			
3. The customer is neutral toward all project vendors.			
4. The customer prefers us.			
How much of the project output is generated by our organization?	2		
1. Less than 50%			
2. 50% to 70%			
3. 70% to 90%			
4. More than 90%			
What is the startup cost of this project?	1		
1. High			
2. Moderate			
3. Low			
4. Minimal			

Table A.2 (continued)

QUESTION	IMPACT	PROBABILITY	(I × P)
What is the project manager's overall opportunity assessment?	1		
1. High			
2. Moderate			
3. Low			
4. Minimal			

The total opportunity and total risk scores should be plotted on the matrix shown in Table A.3 to determine the project's viability. Projects which fall below a diagonal running from (0,0) to (120,120) should be scrutinized heavily before continuing. These projects will require a high degree of management support and risk containment and may offer lower paybacks than those projects that plot above the diagonal.

TABLE A.3   Assessment Matrix

Appendix B:
Comparison of software products

TABLE B.1   Product Pros and Cons

Software Product	Pros	Cons
Artemis Views	Successful integration of MS Project 98	Lack of calculated fields in the standard version
	Good, desktop-level planning functions and resource management	Problems defining custom fields
		Does not manage consumable resources
	Administration console	PERT can only be viewed as a print view
	Data synchronization	
		Euro not included as an alternate currency

TABLE B.1 (continued)

Software Product	Pros	Cons
CA-Super Project	Advanced ergonomics Comprehensive planning functions Calculates efficiency factor for resources Graphical WBS editor Multiproject management Functions available for resolving over allocations Web enabled	No prebuilt interfacing capabilities with ERM software Project data locked at a project file level Can save only three project baselines
Microsoft Project 98	Same interface as the rest of the MS Office family Support for OLE automation Rich detailed desktop-planning functions Includes resource contouring Dynamic tables for use in Excel	Weak management of multiple users and multiple projects Estimation functions are not covered Gaps in resource management Inflexible for time entry
Open Plan Professional	Good ergonomics Complete tutorial Impressive cross tab, view, sort, and multilevel filter functions Excellent planning, optimization, and resource management Manages human, financial, consumable, and perishable resources Web enabled Interfaces with ERM	Limited scalability No code editor No project simulation function No estimation function No graphical report generator

TABLE B.1 (continued)

SOFTWARE PRODUCT	PROS	CONS
Project Scheduler 7	Easy to use wizards	No workload estimation functions
	Ability to generate Gantt chart of resources	Weak timesheet customization
	Unlimited number of custom fields	Weak in cost control
	Formula fields with Boolean conditions and functions	No advanced reporting and analysis functions
	Discrete distribution of resource effort	
	Manages group resources and skills	
Primavera P3E	User-friendly GUI	P3E is English only
	RTF editor	Little integration between P3E and earlier versions of Primavera PM software
	Web enabled	
	Allows users to publish P3E objects via an HTML translator	Consumable resources not managed
	Good resource management features	No skills management
Results Management	Repository administration and workstation deployment functions	Poor cost-management functionality
	Remote data synchronization	Weak Gantt chart generation
	Planning and estimation tool included	Import/export of Excel data not supported
	Excellent resource and allocations management	
PlanView	Easy to use	Financial management and cost-management functionality limited
	Powerful planning functionality	
	Unlimited what-if scenarios	Limited international presence
	Web enabled	
	Strong resource-management functionality	

Source: Software Magazine, April/May 2000, p. 38.

Appendix C:
Case study

As a way of tying all the information in this book together, I will present a case study in this appendix of a project on which I participated which will illustrate some of the do's and don'ts of team formation and project leadership. This particular project was one in which the IT software development group was tasked with designing a new and improved system for processing customer-usage transactions. The old system was a hodgepodge of programs written in different languages in different styles over the years. Band-aid solutions had been applied over the years, but the nature of the customer-usage transactions had changed to the point where the old system was inefficient and could not be effectively updated anymore. The accounting department, being the owner of the system and the associated data, agreed to initiate a project to replace the old usage processing system.

Right away we could see a problem developing. The request to replace the system came from the IT department, rather than the user department. The users felt some pain from using the old system, but the this-is-how-we've-always-done-it mentality prevailed. At the same time the IT department was feeling tremendous pain from trying to keep the old system afloat. It took a great deal of negotiating, employing many of the negotiation strategies from Chapter 5, to convince the accountants to go along with a rewrite project. We had to show them in dollars and cents how we could spend some time and money now to save them time and money later. This is a good thing, to be sure. The problem arises from the fact that since we had to convince accounting to come along, we were faced with lukewarm sponsorship throughout the life of the project. As discussed in Chapters 1 and 2, project sponsorship is of the utmost importance to the success of the project as this is where your access to project resources comes from. More on resource allocations later, but suffice it to say for now, resource allocation was a problem throughout the life of this project.

Once the decision was made to proceed with the project, a project team had to be formed. We were able to form a team comprised of both IT and accounting personnel from various levels of each organization and a project leader, from IT, was appointed to lead the team. Again, we can see a problem from the start with the appointment of the project leader. I have often seen IT personnel chosen to lead projects simply because they had the project management skills. However, if the IT (or any other high-tech group) does not a have a main stake in a particular project, it is much better to pull the project lead either from the sponsoring department or from a project management office. It is difficult to build your power base, as discussed in Chapters 10 and 11, if you are not closely connected to the project sponsors or if you don't have some authority behind you such as that from a project management office.

The first order of business for the team was to compile requirements for the new system. Guidelines as found in Chapter 2 were more or less followed and the requirements definition phase of the project actually progressed pretty well. The major consumer of system output, accounting, knew exactly what they wanted in a new system, and the IT group knew exactly what technical specs needed to be included to make the system more manageable. Where the team failed at this step was in interviewing other less obvious stakeholders in the system. Other user departments needed output from the system and the computer operations personnel who were expected to run the system on a daily basis were neither interviewed nor consulted for input into requirements.

As the project progressed, the team went through the normal stages of team development as discussed in Chapter 4. We stormed longer than we should have due to the situations noted above and some problems with our communication plan, or lack thereof (Chapter 6), and problems with the way we managed time on tasks (Chapter 13). We did, however, overcome these difficulties to eventually produce a viable product for our clients. We delayed implementation once for two months to iron out some quality issues (Chapter 8), but once we were implemented everything ran without a hitch. The accountants were actually able, through attrition, to replace two permanent employees with a temporary employee due to improvements made in processing and reporting. My point with this information is to show that despite the difficulties we encountered, and despite the mistakes we made along the way, we had a dedicated, knowledgeable group of team members who persisted until the job was completed properly. If you take nothing else away from this text, remember this: there is no substitute for competent, motivated, project team members. Choose your team members wisely and then do everything in your power to keep them challenged, motivated, compensated, and happy. Do this and your job as a project manager and team leader will be infinitely easier.

About the Author

James Williams received his M.S. degree in project management and has worked as a software engineer and consultant. He has 15 years of experience in the telecommunications industry as both a participant and leader of project teams.

Index

Reengineering Yourself and Your Company: From Engineer to Manager to Leader, Howard Eisner

Successful Marketing Strategy for High-Tech Firms, Second Edition, Eric Viardot

Successful Proposal Strategies for Small Businesses: Winning Government, Private Sector, and International Contracts, Second Edition, Robert S. Frey

Systems Engineering Principles and Practice, H. Robert Westerman

Team Development for High-Tech Project Managers, James Williams

For further information on these and other Artech House titles, including previously considered out-of-print books now available through our In-Print-Forever® (IPF®) program, contact:

Artech House
685 Canton Street
Norwood, MA 02062
Phone: 781-769-9750
Fax: 781-769-6334
e-mail: artech@artechhouse.com

Artech House
46 Gillingham Street
London SW1V 1AH UK
Phone: +44 (0)20 7596-8750
Fax: +44 (0)20 7630-0166
e-mail: artech-uk@artechhouse.com

Find us on the World Wide Web at:
www.artechhouse.com